GRAPHIC HISTORY

THE FIRST MOON LANDING

by Thomas K. Adamson

illustrated by Gordon Purcell and Terry Beatty

Consultant:
Roger D. Launius, PhD
Chair, Division of Space History
National Air and Space Museum
Smithsonian Institution
Washington, D.C.

Capstone press®

Mankato, Minnesota

Graphic Library is published by Capstone Press,
151 Good Counsel Drive, P.O. Box 669, Mankato, Minnesota 56002.
www.capstonepress.com

Printed in the United States of America

1 2 3 4 5 6 11 10 09 08 07 06

Library of Congress Cataloging-in-Publication Data
Adamson, Thomas K., 1970–
The first moon landing / by Thomas K. Adamson; illustrated by Gordon Purcell and Terry Beatty.
p. cm.—(Graphic library. Graphic history)
Includes bibliographical references and index.
ISBN-13: 978-0-7368-6492-3 (hardcover)
ISBN-10: 0-7368-6492-X (hardcover)
ISBN-13: 978-0-7368-7527-1 (softcover pbk.)
ISBN-10: 0-7368-7527-1 (softcover pbk.)
1. Project Apollo (U.S.)—History. 2. Space flight to the moon—History—20th century. 3. Moon—Exploration—20th century. I. Title.
TL789.8.U6A475 2007
629.45'4—dc22 2006008308

Summary: In graphic novel format, tells the story of the *Apollo 11* mission, including the first moon landing in 1969.

Designer
Bob Lentz

Colorist
Otha Zackariah Edward Lohse

Editor
Donald Lemke

Photo credit: Shutterstock/Taipan Kid, 29

Editor's note: Direct quotations from primary sources are indicated by a yellow background.

Direct quotations appear on the following pages:
Page 5, from President John F. Kennedy's *Special Message to the Congress on Urgent National Needs* on May 25, 1961, as transcribed at the John F. Kennedy Library and Museum Online (http://www.jfklibrary.org/Historical+Resources/Archives/Reference+Desk/Speeches/JFK/Urgent+National+Needs+Page+4.htm).
Pages 10, 12, 13, 14, 16, 17, 18, 20, 21, 22, from *Apollo 11 PAO Mission Commentary Transcript* (http://www.jsc.nasa.gov/history/mission_trans/apollo11.htm).

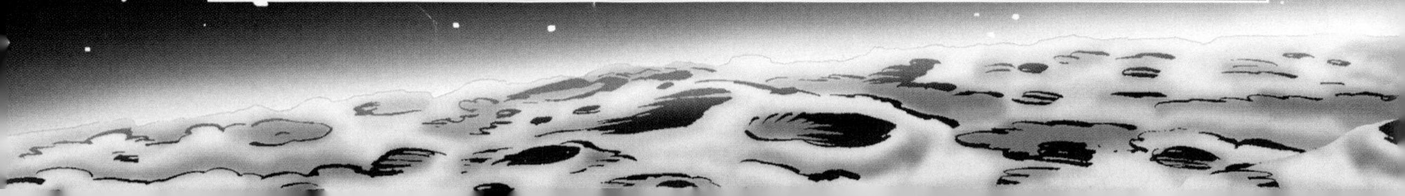

TABLE OF CONTENTS

CHAPTER 1
THE SPACE RACE

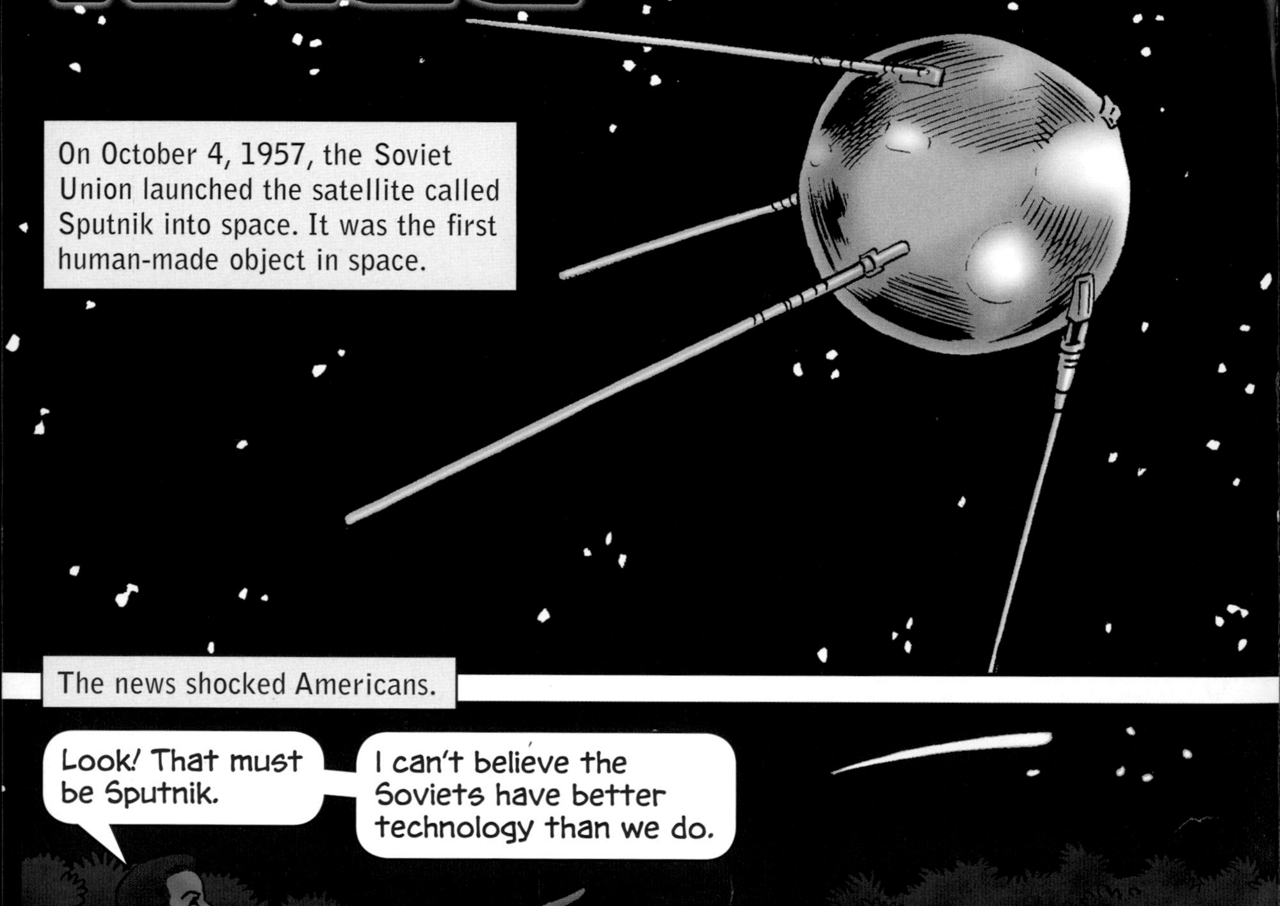
On October 4, 1957, the Soviet Union launched the satellite called Sputnik into space. It was the first human-made object in space.

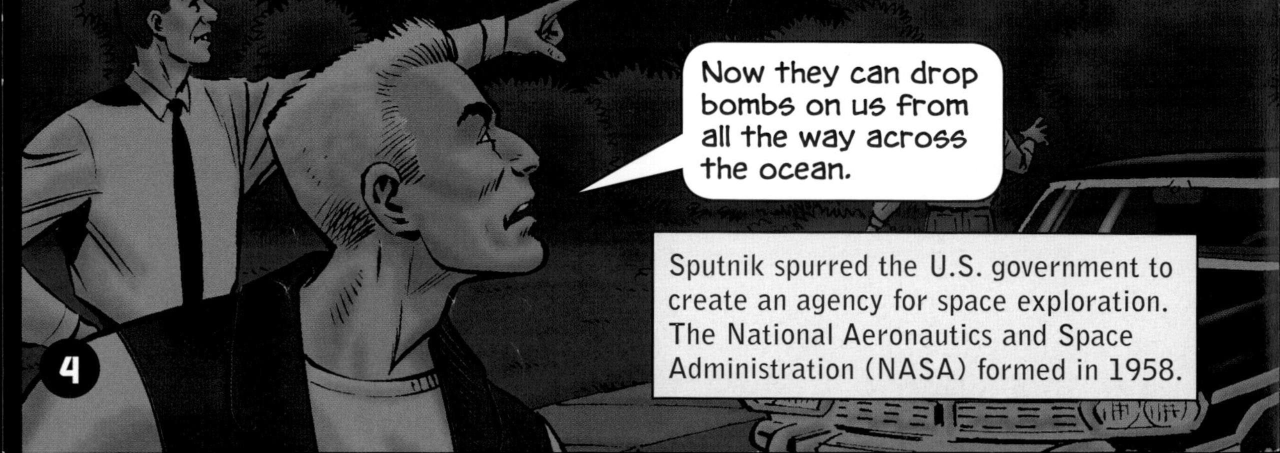
The news shocked Americans.
Look! That must be Sputnik.
I can't believe the Soviets have better technology than we do.
Now they can drop bombs on us from all the way across the ocean.
Sputnik spurred the U.S. government to create an agency for space exploration. The National Aeronautics and Space Administration (NASA) formed in 1958.

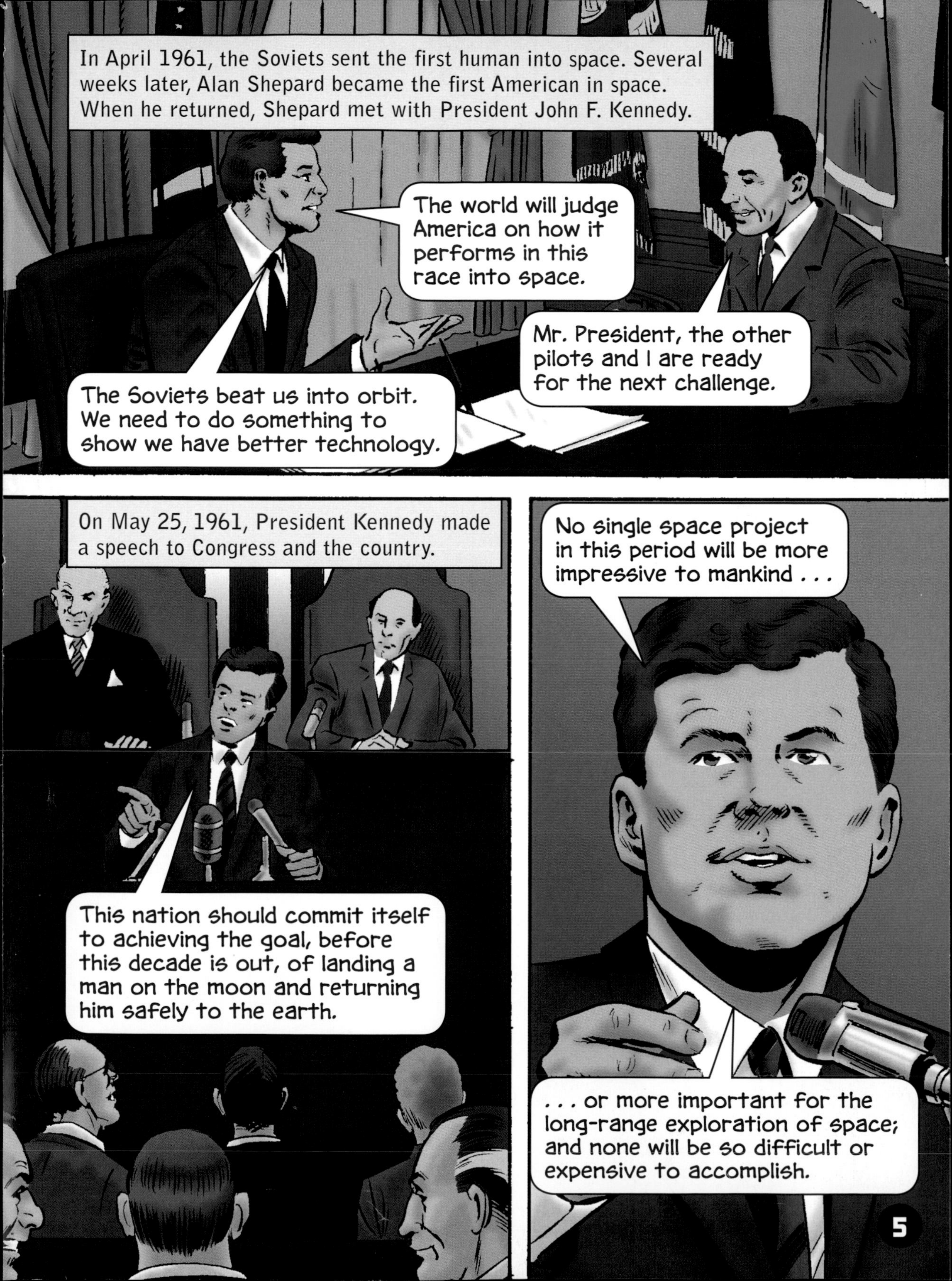
In April 1961, the Soviets sent the first human into space. Several weeks later, Alan Shepard became the first American in space. When he returned, Shepard met with President John F. Kennedy.
The world will judge America on how it performs in this race into space.
Mr. President, the other pilots and I are ready for the next challenge.
The Soviets beat us into orbit. We need to do something to show we have better technology.
On May 25, 1961, President Kennedy made a speech to Congress and the country.
This nation should commit itself to achieving the goal, before this decade is out, of landing a man on the moon and returning him safely to the earth.
No single space project in this period will be more impressive to mankind . . .
. . . or more important for the long-range exploration of space; and none will be so difficult or expensive to accomplish.

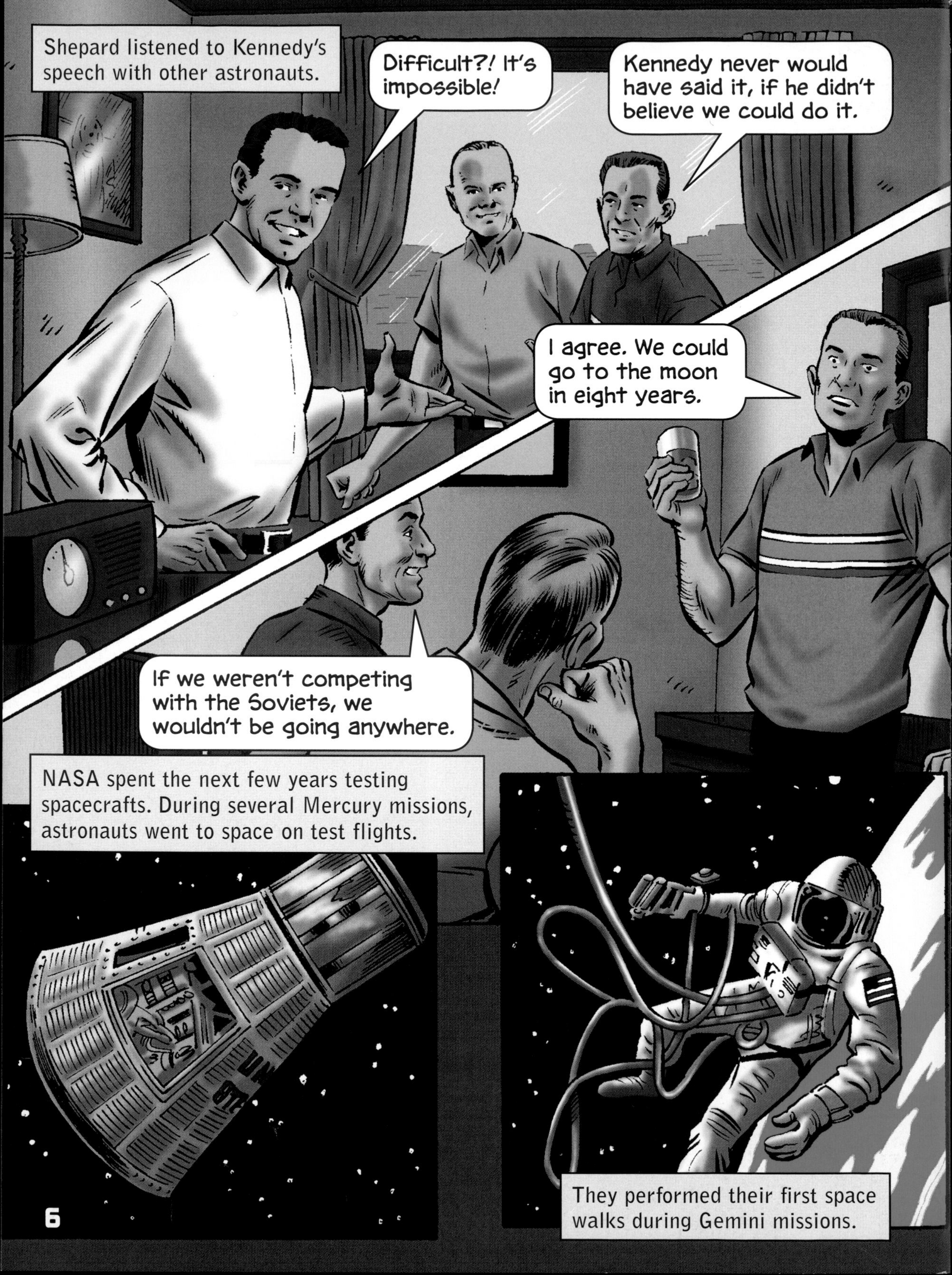
Shepard listened to Kennedy's speech with other astronauts.
Difficult?! It's impossible!
Kennedy never would have said it, if he didn't believe we could do it.
I agree. We could go to the moon in eight years.
If we weren't competing with the Soviets, we wouldn't be going anywhere.
NASA spent the next few years testing spacecrafts. During several Mercury missions, astronauts went to space on test flights.
They performed their first space walks during Gemini missions.

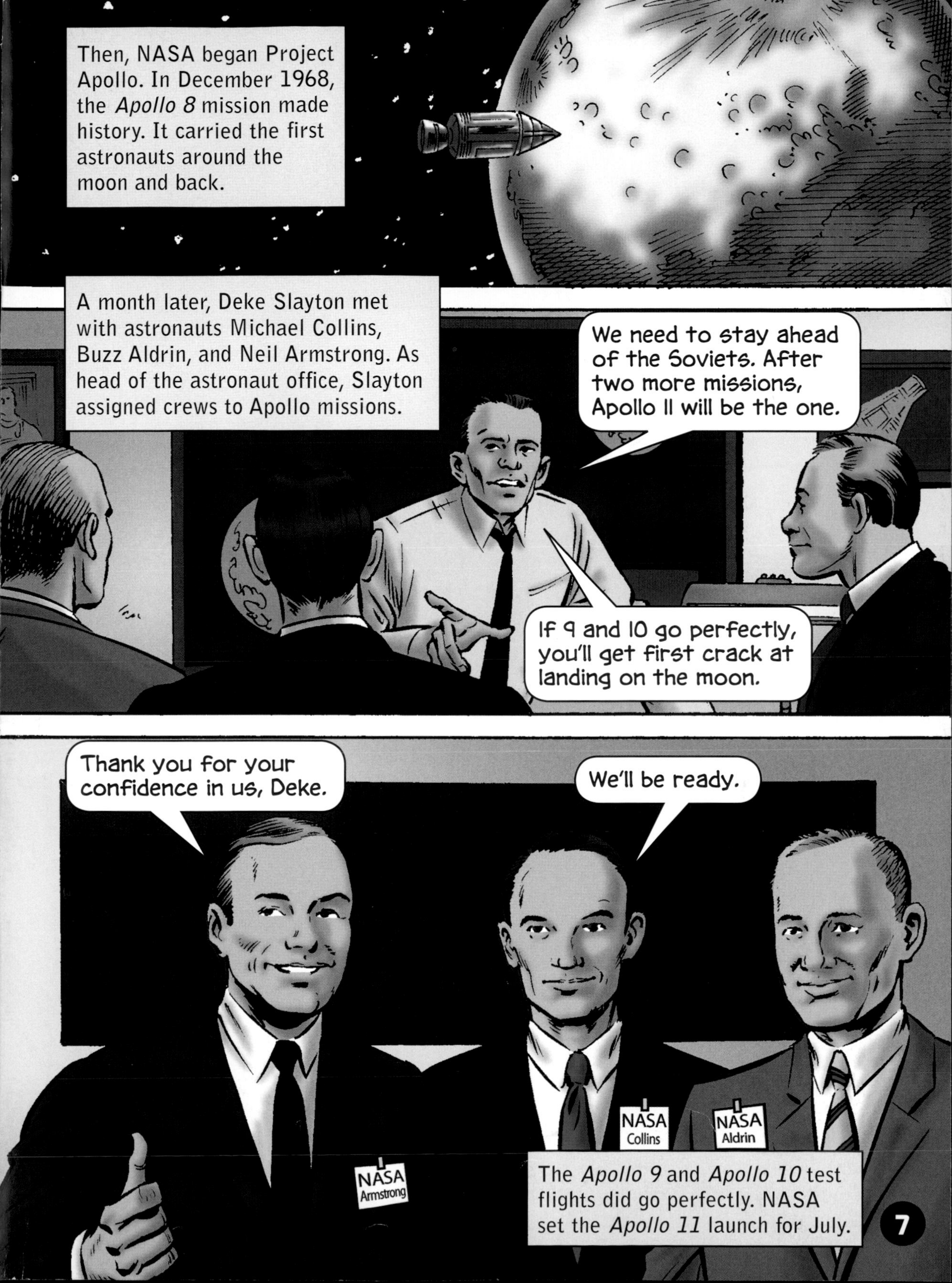
Then, NASA began Project Apollo. In December 1968, the *Apollo 8* mission made history. It carried the first astronauts around the moon and back.
A month later, Deke Slayton met with astronauts Michael Collins, Buzz Aldrin, and Neil Armstrong. As head of the astronaut office, Slayton assigned crews to Apollo missions.
We need to stay ahead of the Soviets. After two more missions, Apollo 11 will be the one.
If 9 and 10 go perfectly, you'll get first crack at landing on the moon.
Thank you for your confidence in us, Deke.
We'll be ready.
NASA Armstrong
NASA Collins
NASA Aldrin
The *Apollo 9* and *Apollo 10* test flights did go perfectly. NASA set the *Apollo 11* launch for July.

CHAPTER 2
TO THE MOON

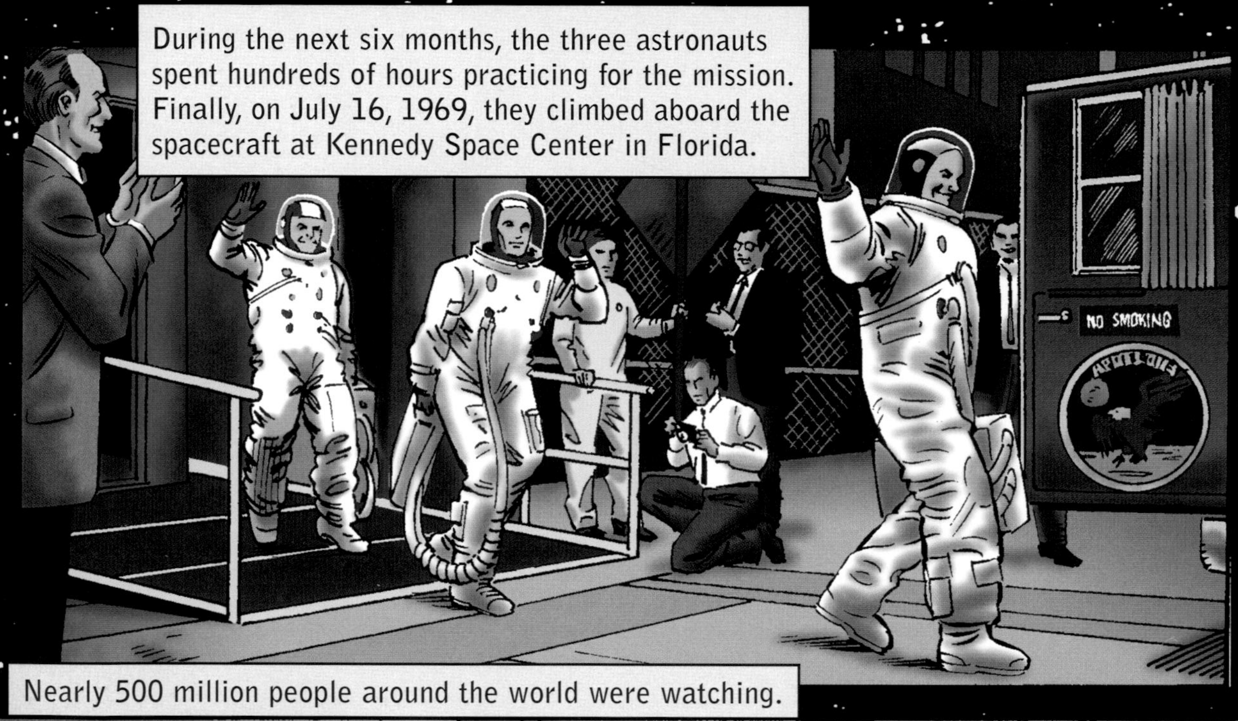

During the next six months, the three astronauts spent hundreds of hours practicing for the mission. Finally, on July 16, 1969, they climbed aboard the spacecraft at Kennedy Space Center in Florida.
NO SMOKING
Nearly 500 million people around the world were watching.

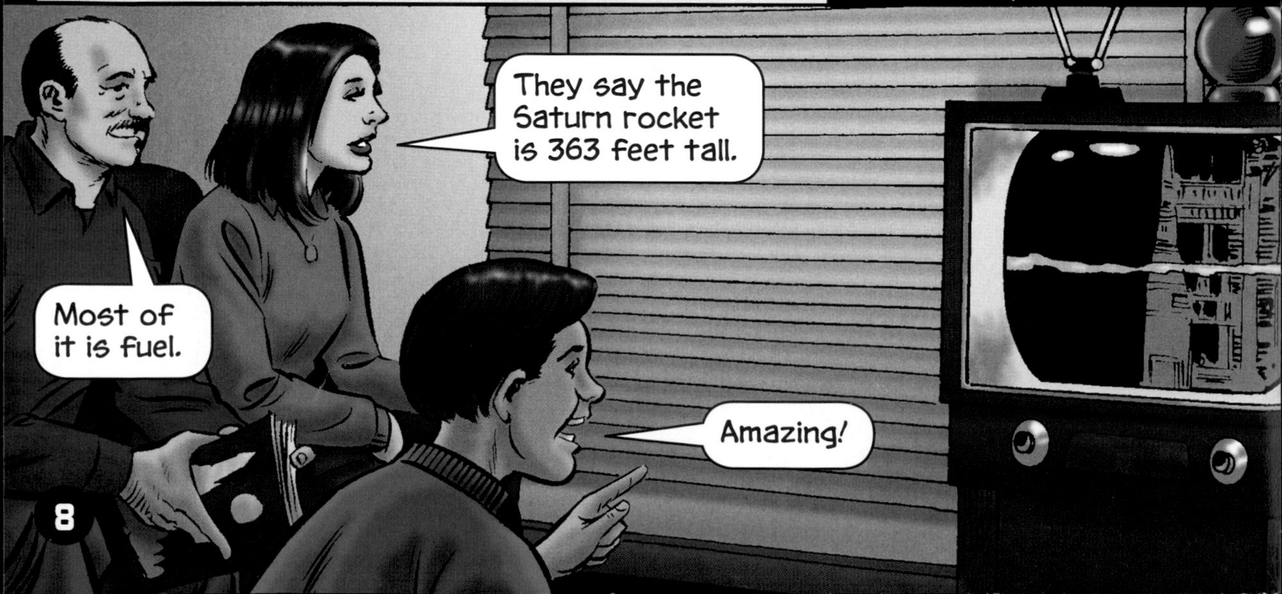

They say the Saturn rocket is 363 feet tall.
Most of it is fuel.
Amazing!

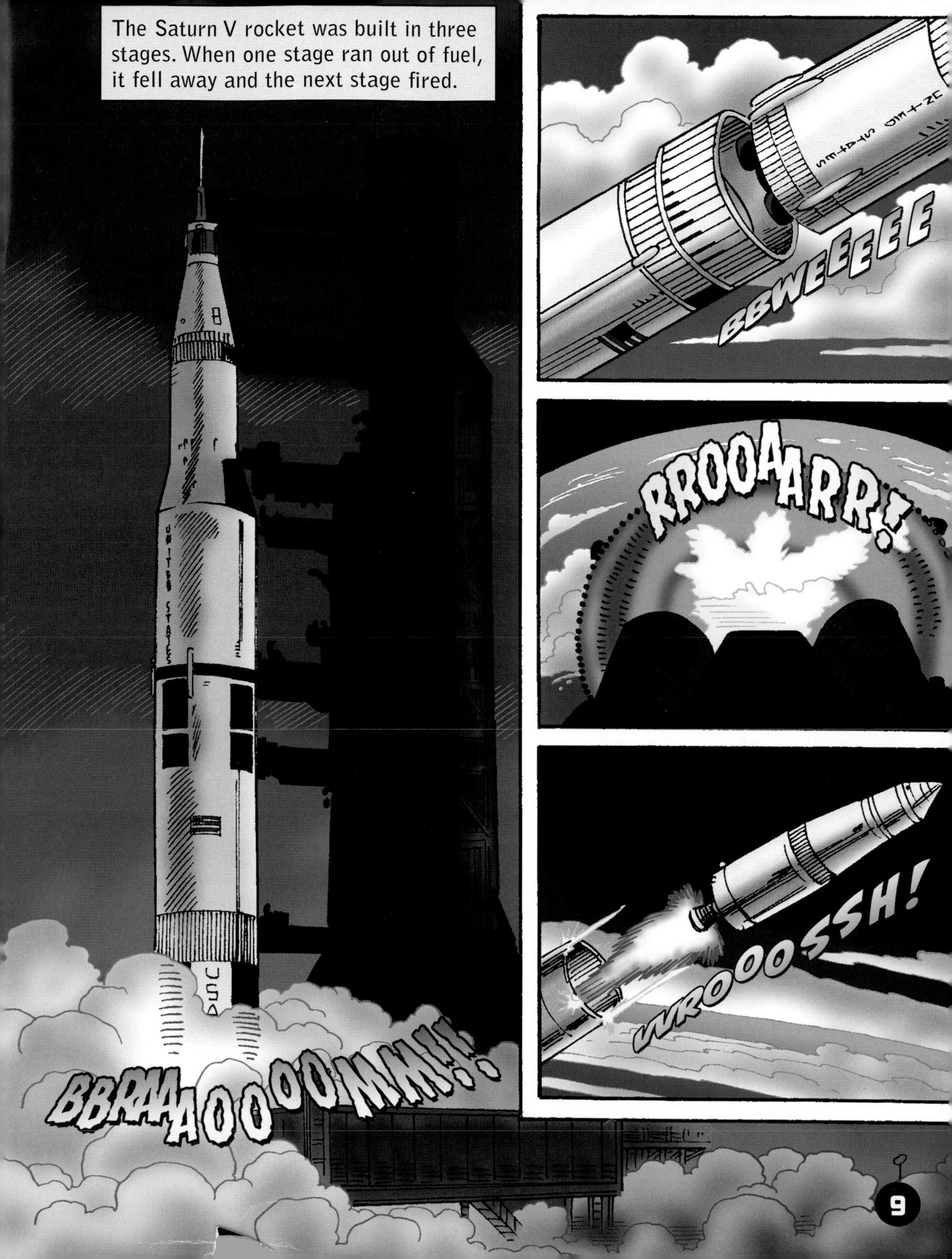
The Saturn V rocket was built in three stages. When one stage ran out of fuel, it fell away and the next stage fired.
BBRAAAOOOOMMM!!!
BBWEEEEE
RROOAARR!
WROOOSSH!

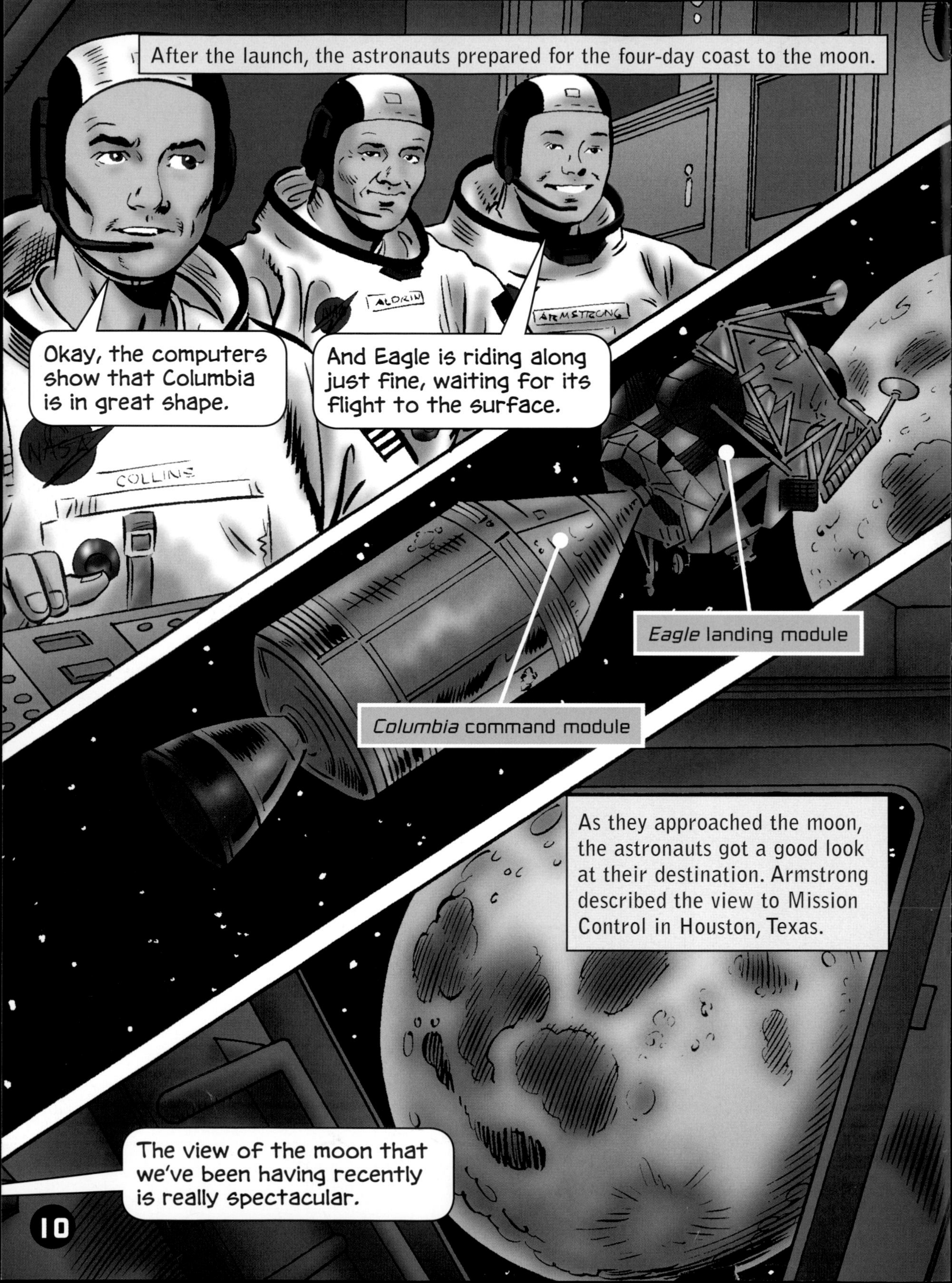
After the launch, the astronauts prepared for the four-day coast to the moon.
Okay, the computers show that Columbia is in great shape.
And Eagle is riding along just fine, waiting for its flight to the surface.
Eagle landing module
Columbia command module
As they approached the moon, the astronauts got a good look at their destination. Armstrong described the view to Mission Control in Houston, Texas.
The view of the moon that we've been having recently is really spectacular.

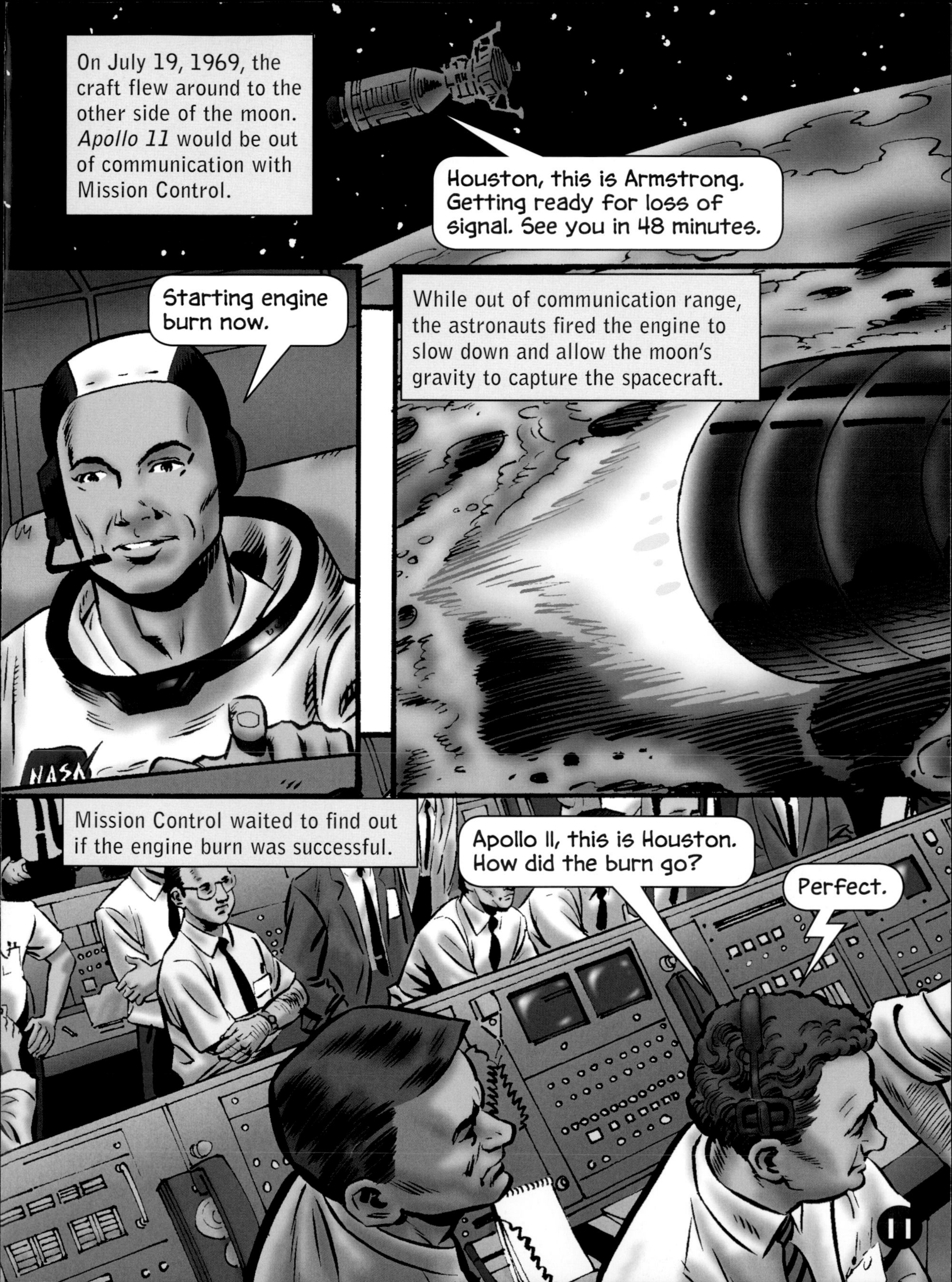
On July 19, 1969, the craft flew around to the other side of the moon. *Apollo 11* would be out of communication with Mission Control.
Houston, this is Armstrong. Getting ready for loss of signal. See you in 48 minutes.
Starting engine burn now.
While out of communication range, the astronauts fired the engine to slow down and allow the moon's gravity to capture the spacecraft.
NASA
Mission Control waited to find out if the engine burn was successful.
Apollo 11, this is Houston. How did the burn go?
Perfect.

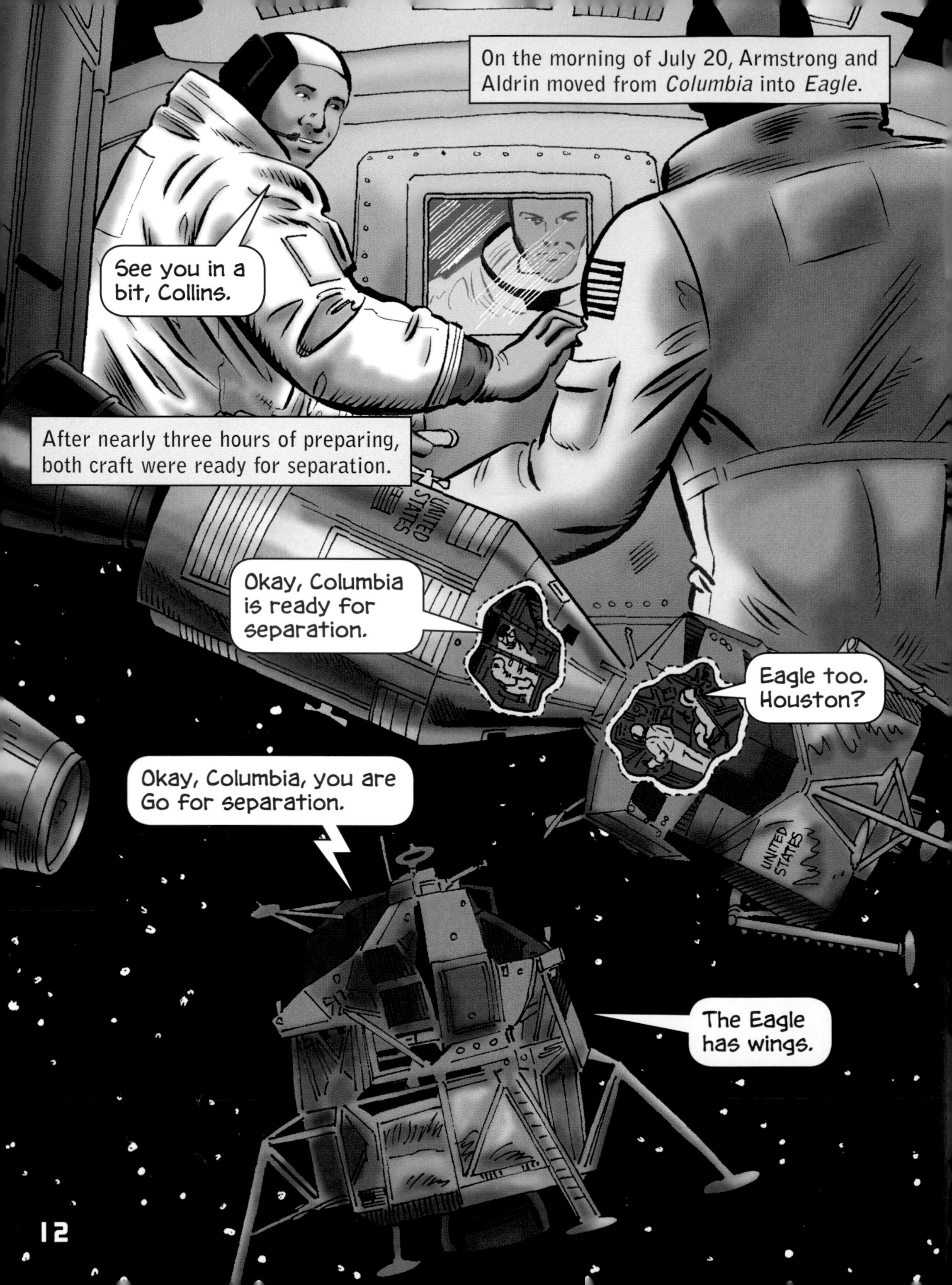

On the morning of July 20, Armstrong and Aldrin moved from *Columbia* into *Eagle*.
See you in a bit, Collins.
After nearly three hours of preparing, both craft were ready for separation.
UNITED STATES
Okay, Columbia is ready for separation.
Eagle too. Houston?
Okay, Columbia, you are Go for separation.
UNITED STATES
The Eagle has wings.

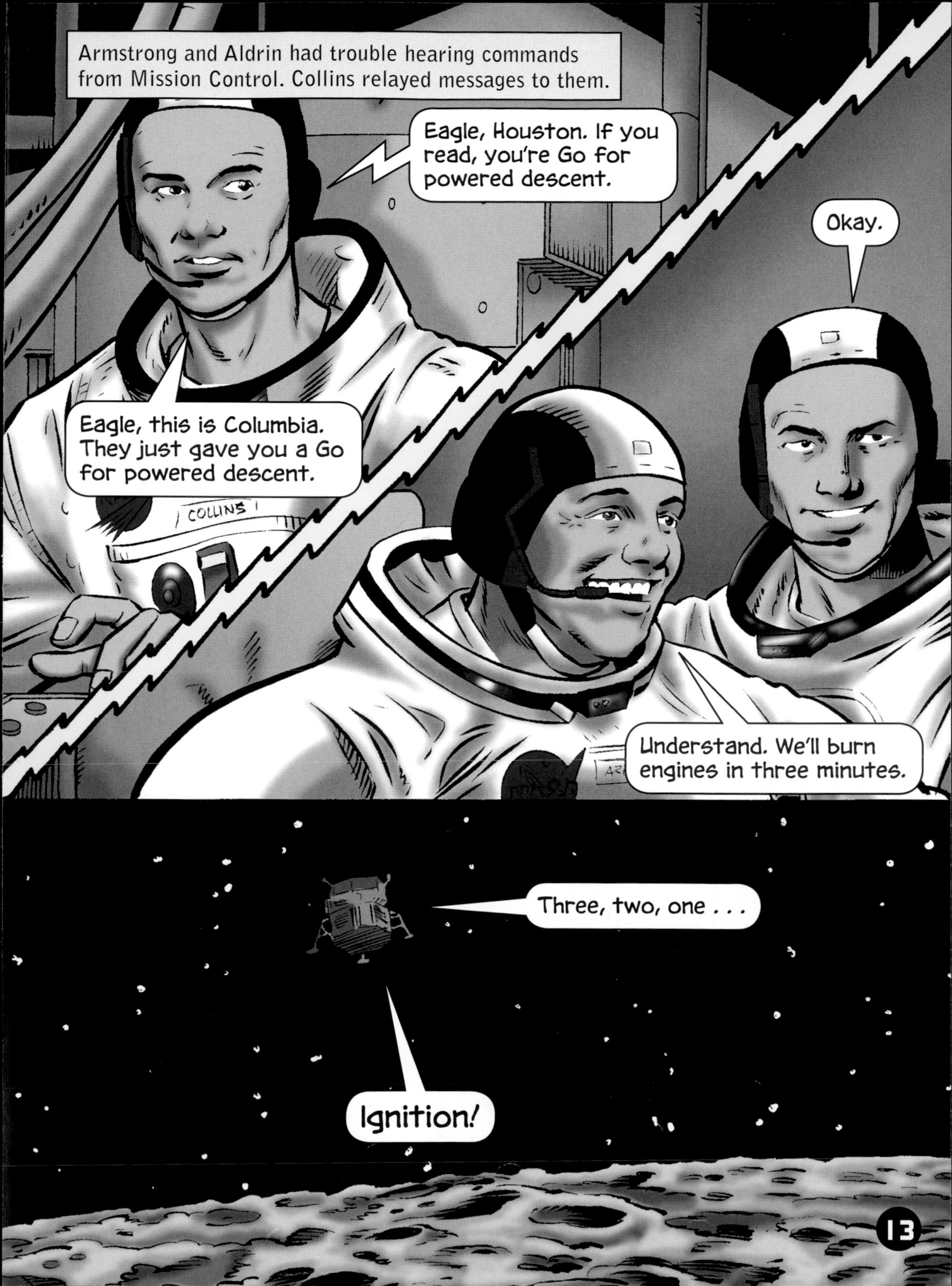
Armstrong and Aldrin had trouble hearing commands from Mission Control. Collins relayed messages to them.
Eagle, Houston. If you read, you're Go for powered descent.
Eagle, this is Columbia. They just gave you a Go for powered descent.
COLLINS
Okay.
Understand. We'll burn engines in three minutes.
Three, two, one . . .
Ignition!

CHAPTER 3
ONE SMALL STEP

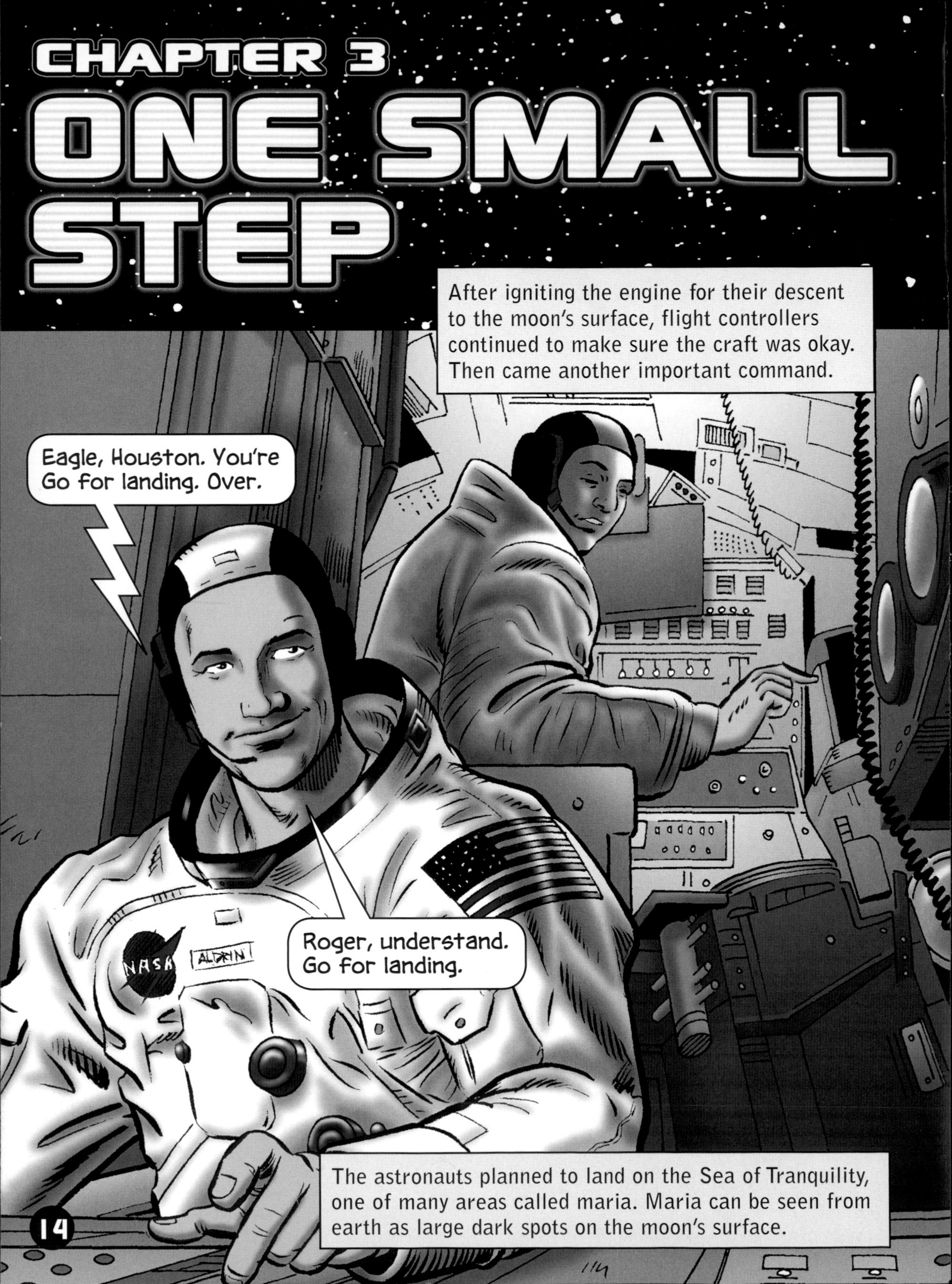
After igniting the engine for their descent to the moon's surface, flight controllers continued to make sure the craft was okay. Then came another important command.
Eagle, Houston. You're Go for landing. Over.
Roger, understand. Go for landing.
NASA
ALDRIN
The astronauts planned to land on the Sea of Tranquility, one of many areas called maria. Maria can be seen from earth as large dark spots on the moon's surface.

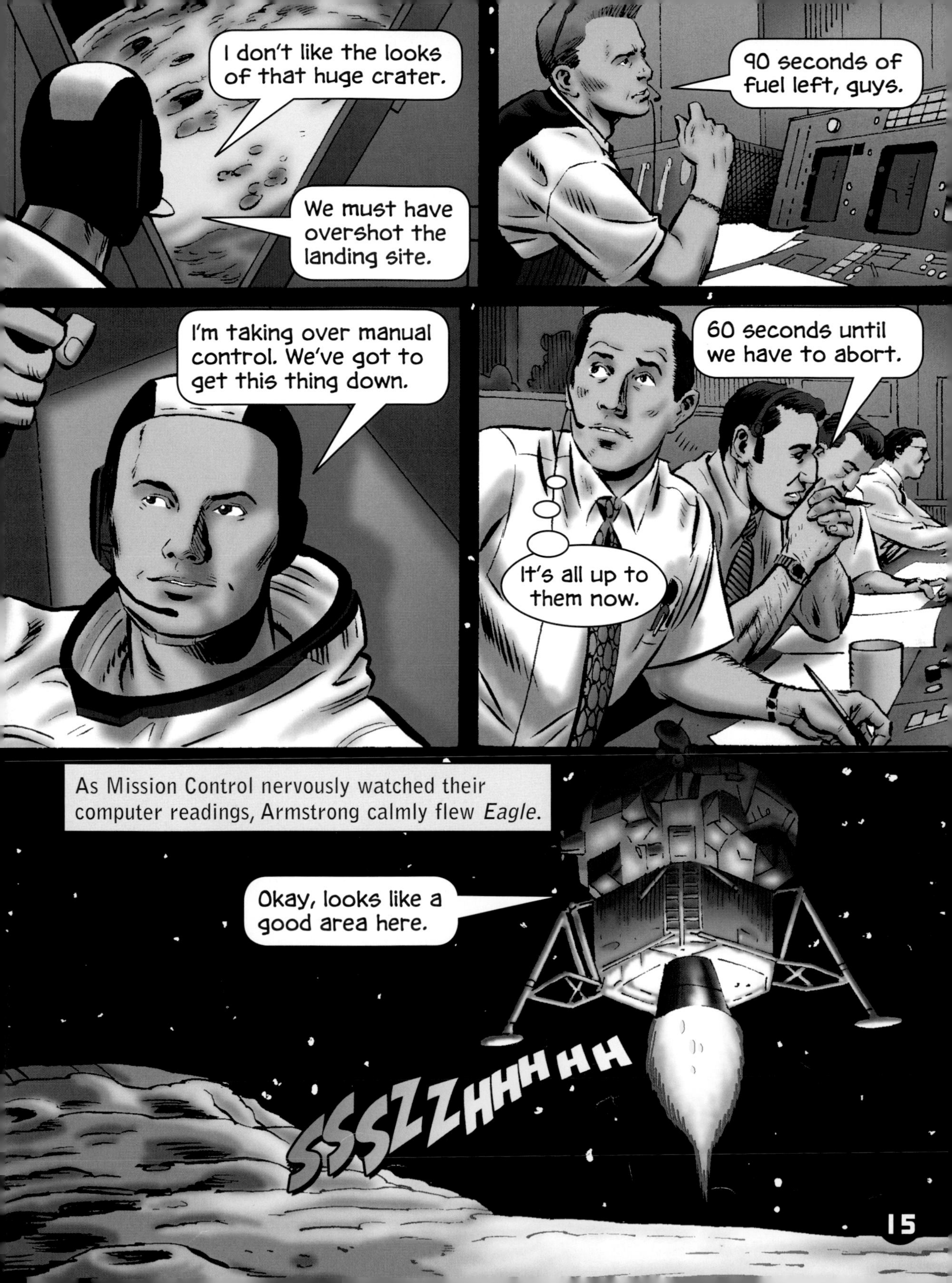
I don't like the looks of that huge crater.
We must have overshot the landing site.
90 seconds of fuel left, guys.
I'm taking over manual control. We've got to get this thing down.
60 seconds until we have to abort.
It's all up to them now.
As Mission Control nervously watched their computer readings, Armstrong calmly flew *Eagle*.
Okay, looks like a good area here.
SSSZZHHHHHH

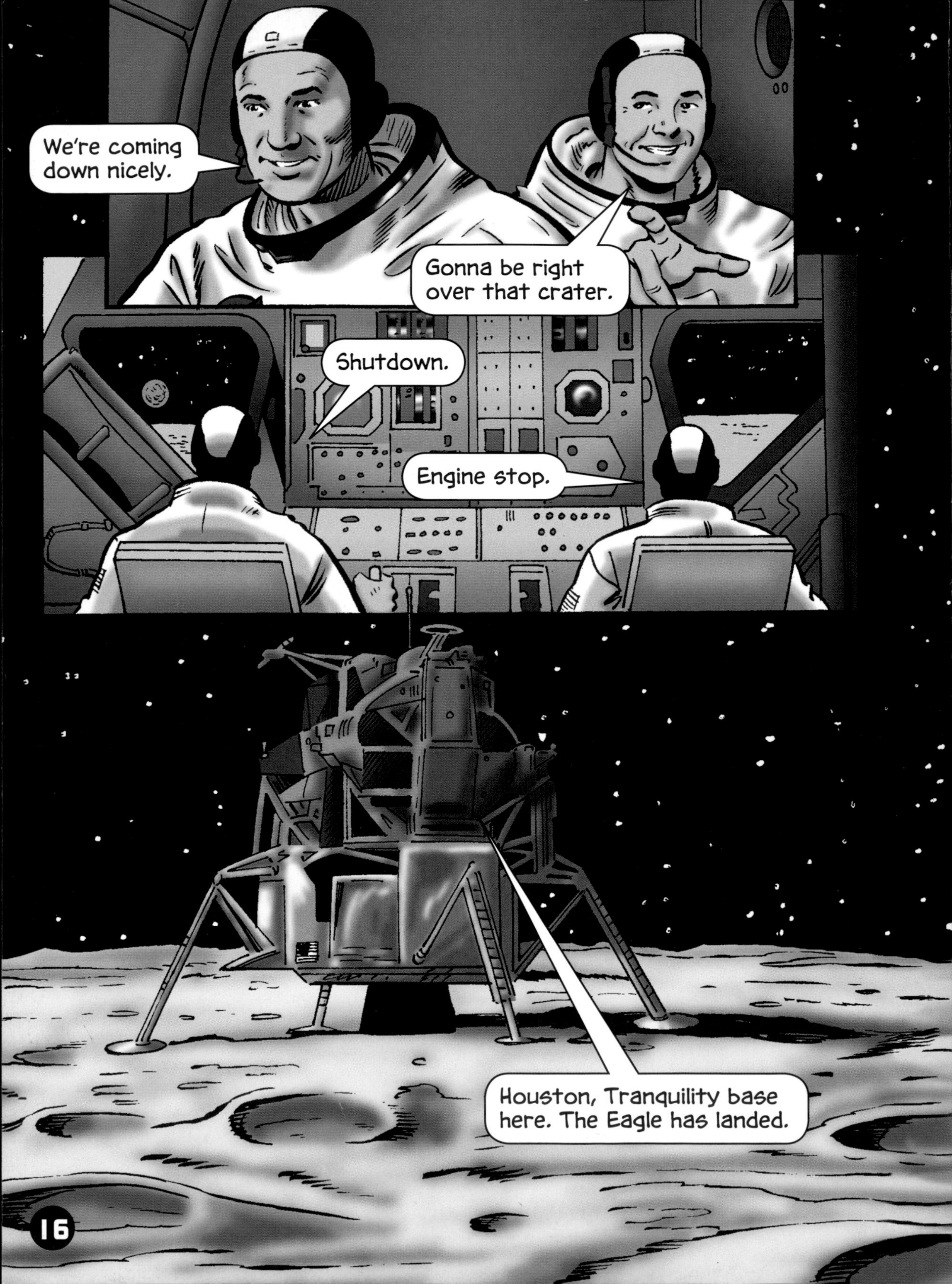
We're coming down nicely.
Gonna be right over that crater.
Shutdown.
Engine stop.
Houston, Tranquility base here. The Eagle has landed.

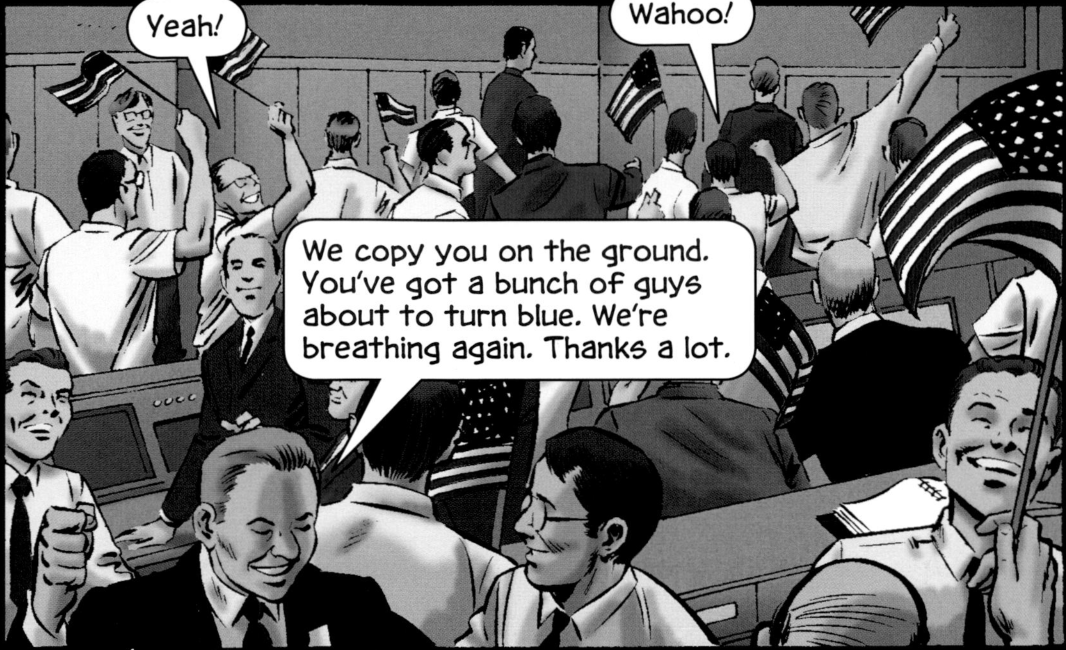
Yeah!
Wahoo!
We copy you on the ground. You've got a bunch of guys about to turn blue. We're breathing again. Thanks a lot.

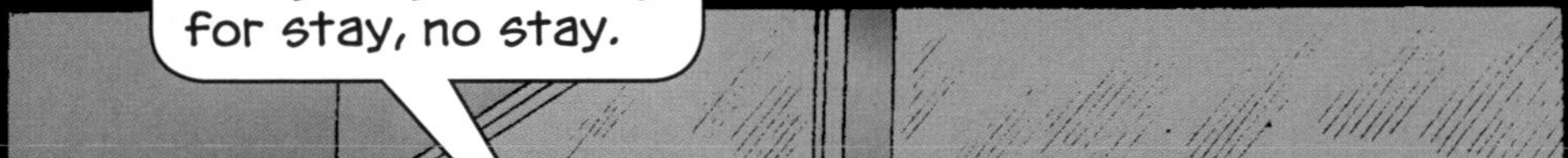
All right everyone, calm down.
Everybody stand by for stay, no stay.

There was little time to celebrate. Flight controllers had only one minute to make sure the spacecraft was okay. If anything was wrong, the craft would have to lift off immediately.

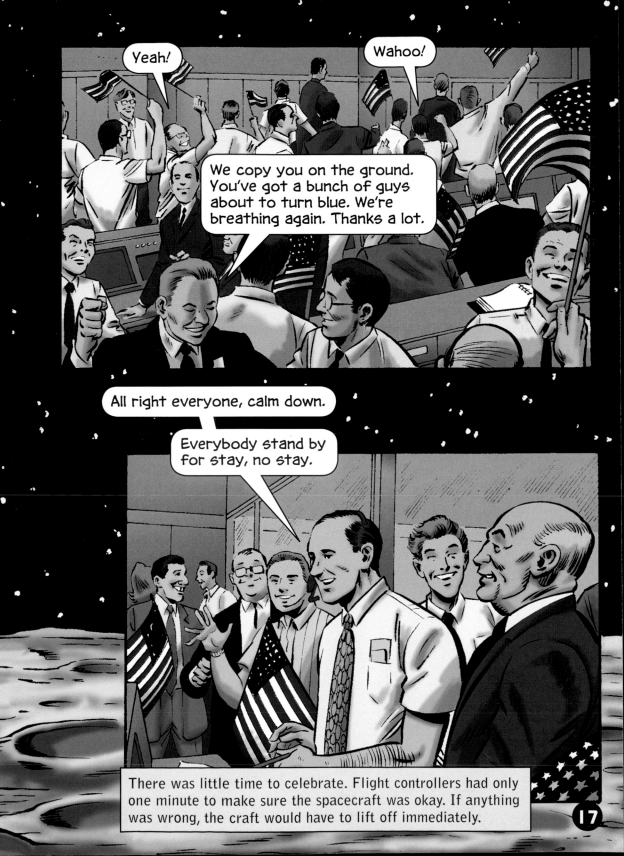

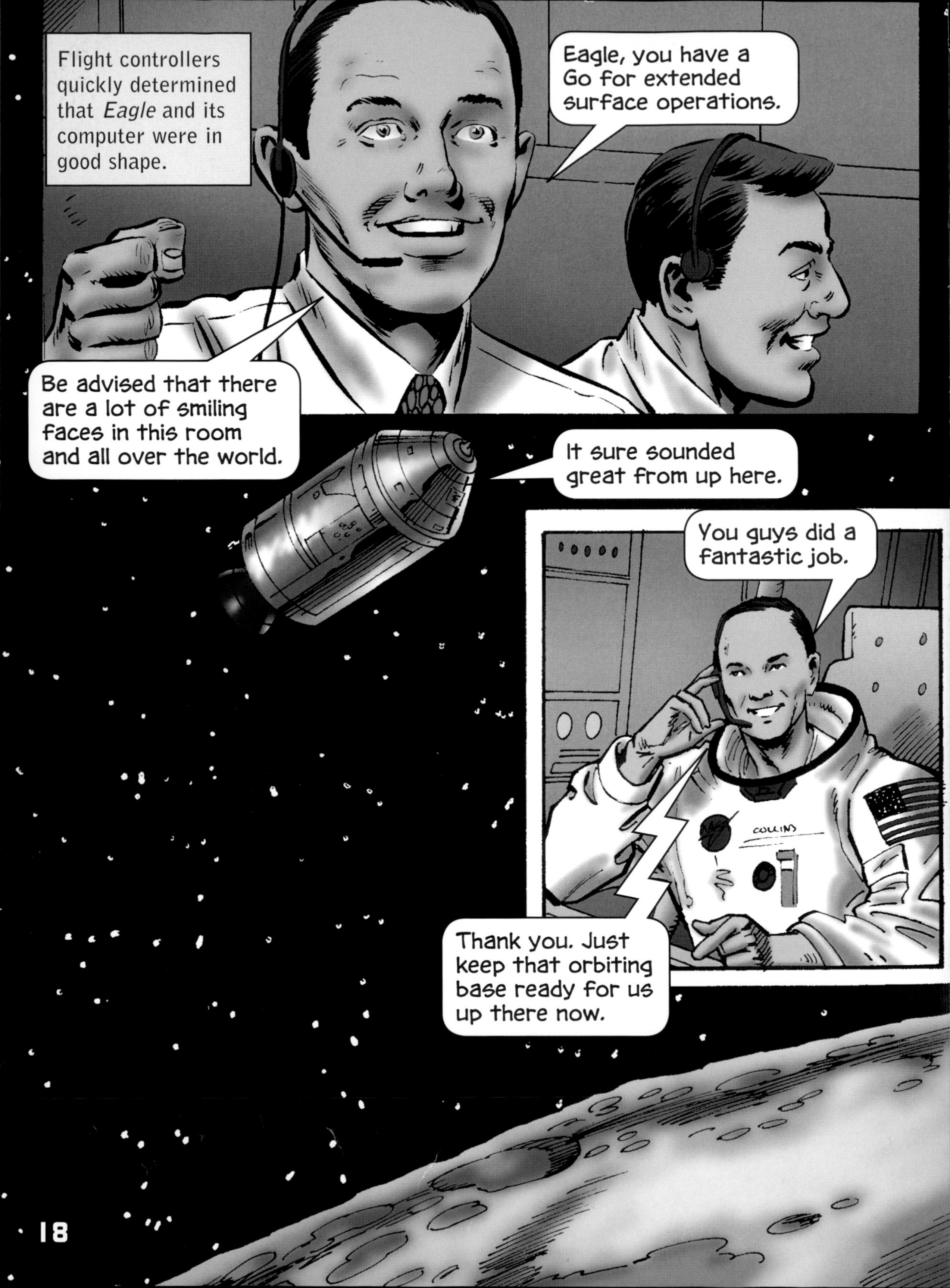
Flight controllers quickly determined that *Eagle* and its computer were in good shape.
Eagle, you have a Go for extended surface operations.
Be advised that there are a lot of smiling faces in this room and all over the world.
It sure sounded great from up here.
You guys did a fantastic job.
Thank you. Just keep that orbiting base ready for us up there now.
COLLINS

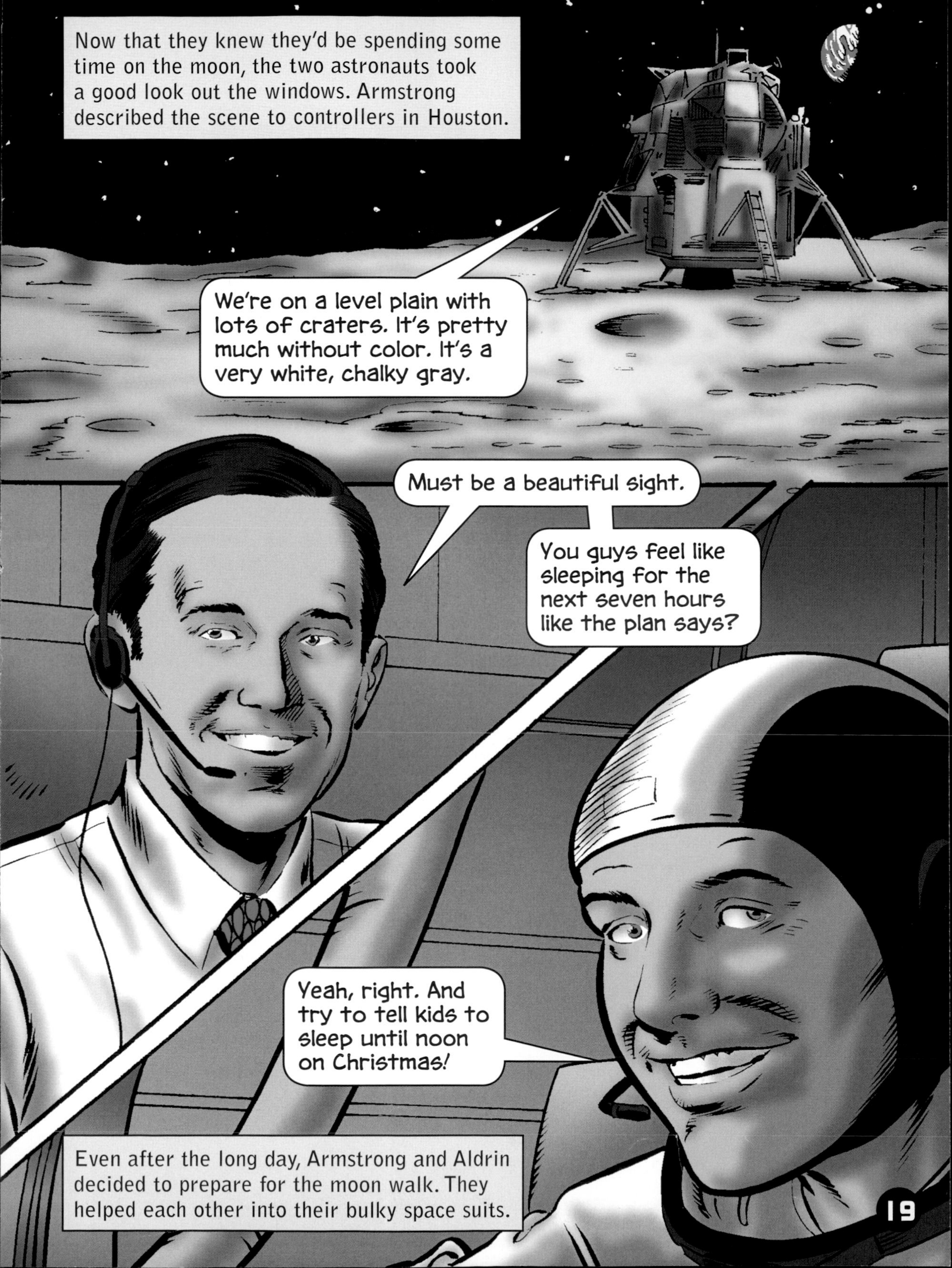
Now that they knew they'd be spending some time on the moon, the two astronauts took a good look out the windows. Armstrong described the scene to controllers in Houston.
We're on a level plain with lots of craters. It's pretty much without color. It's a very white, chalky gray.
Must be a beautiful sight.
You guys feel like sleeping for the next seven hours like the plan says?
Yeah, right. And try to tell kids to sleep until noon on Christmas!
Even after the long day, Armstrong and Aldrin decided to prepare for the moon walk. They helped each other into their bulky space suits.

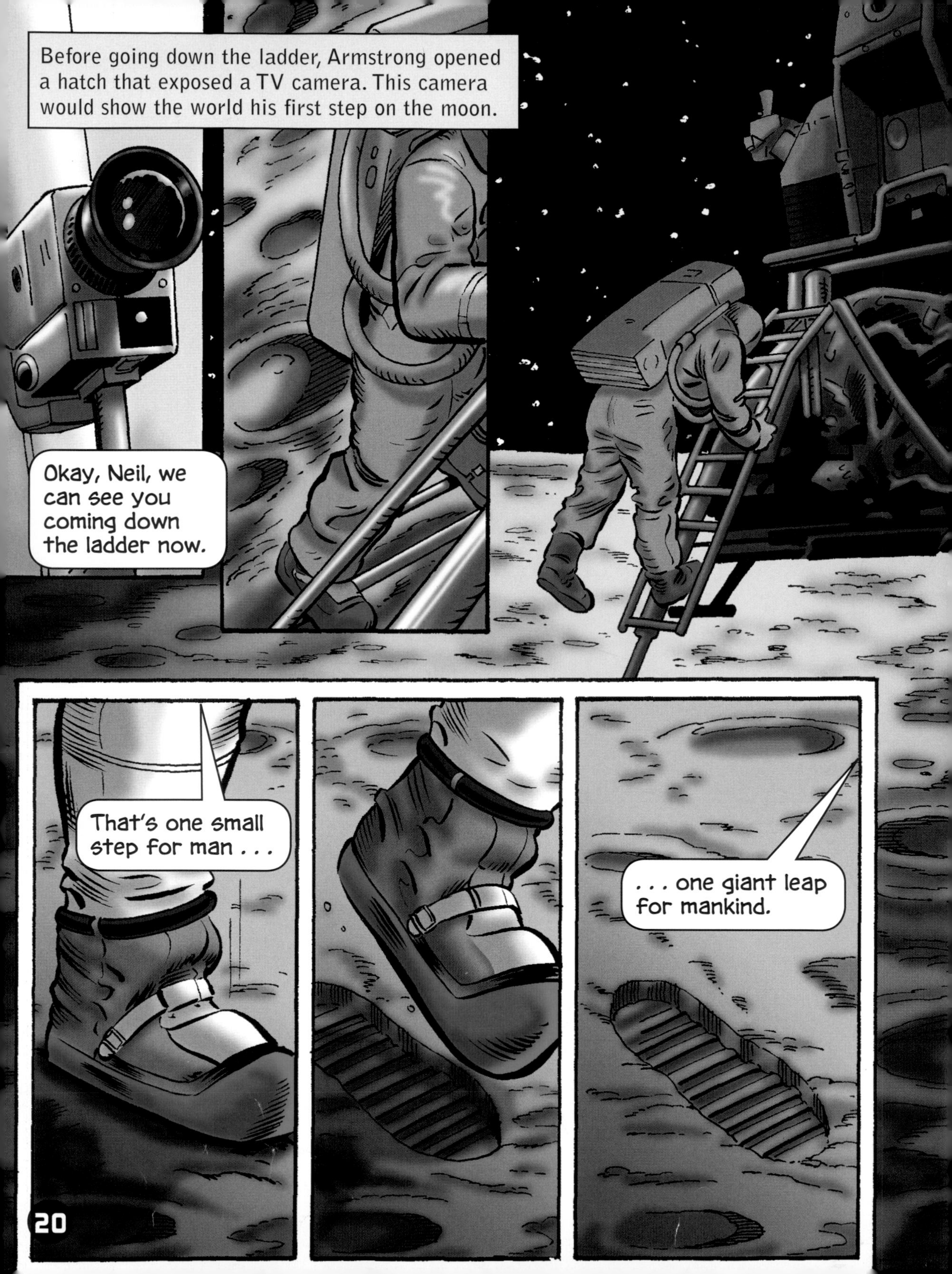
Before going down the ladder, Armstrong opened a hatch that exposed a TV camera. This camera would show the world his first step on the moon.
Okay, Neil, we can see you coming down the ladder now.
That's one small step for man . . .
. . . one giant leap for mankind.

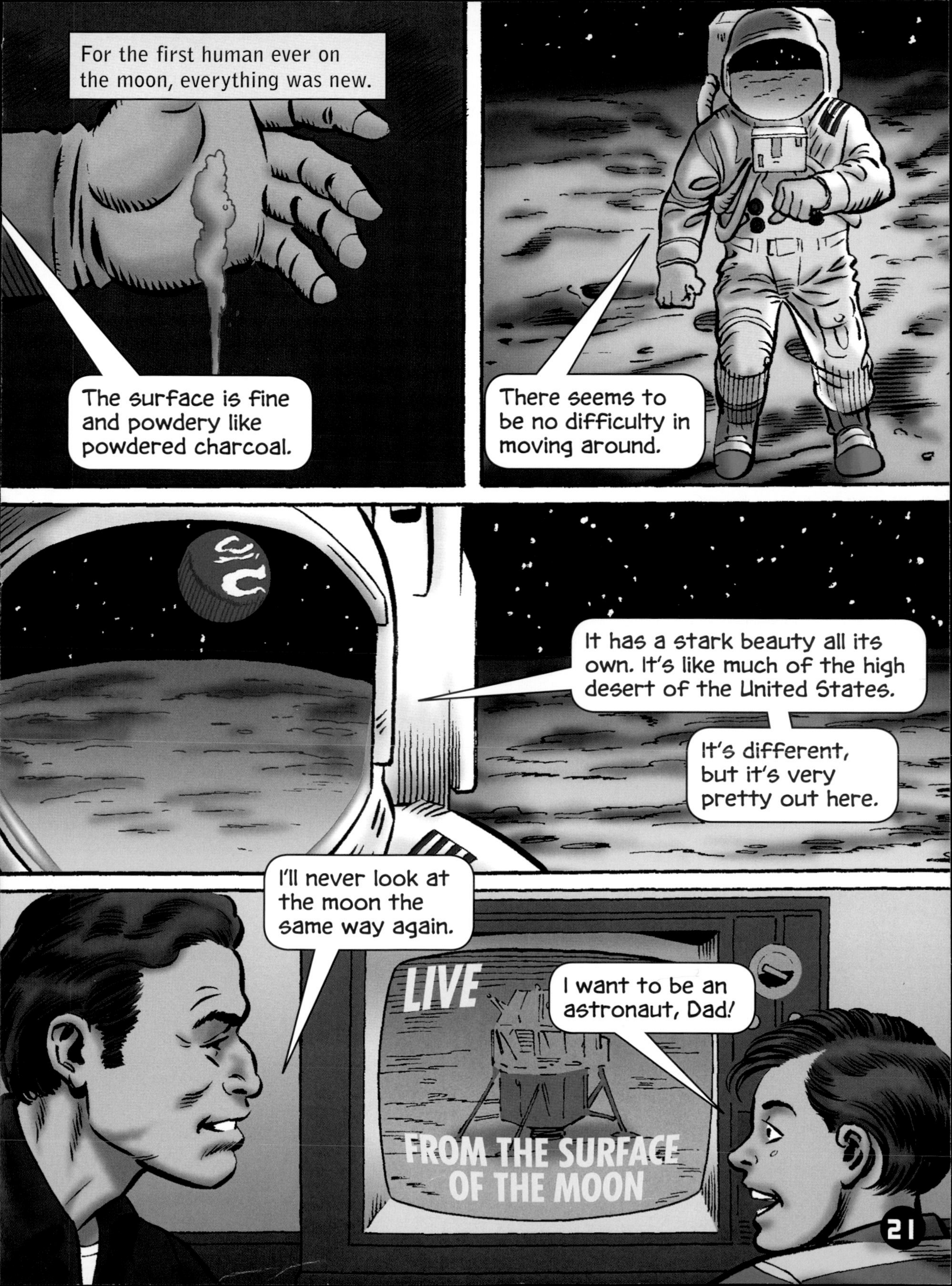
For the first human ever on the moon, everything was new.
The surface is fine and powdery like powdered charcoal.
There seems to be no difficulty in moving around.
It has a stark beauty all its own. It's like much of the high desert of the United States.
It's different, but it's very pretty out here.
I'll never look at the moon the same way again.
LIVE
FROM THE SURFACE OF THE MOON
I want to be an astronaut, Dad!

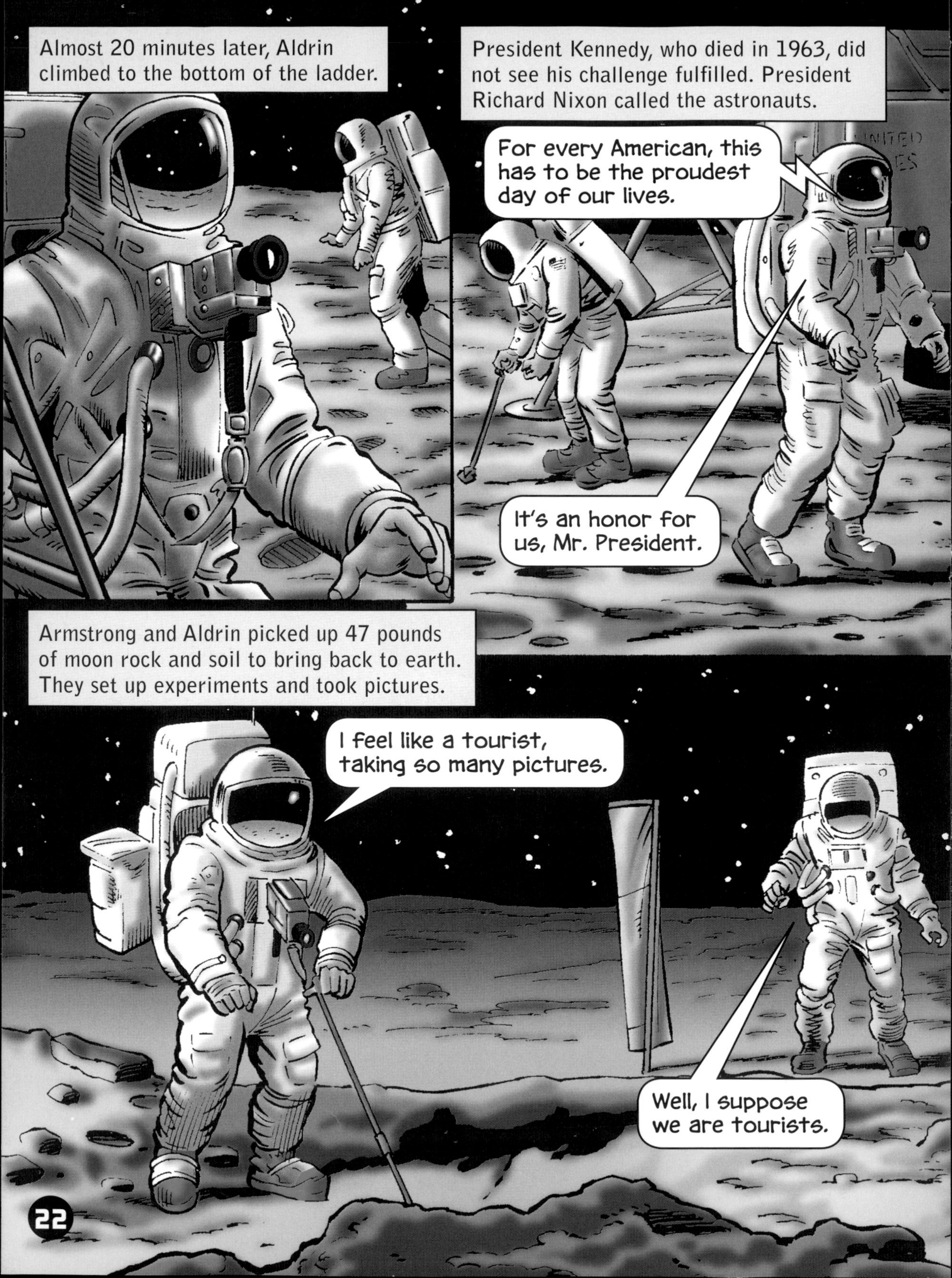
Almost 20 minutes later, Aldrin climbed to the bottom of the ladder.
President Kennedy, who died in 1963, did not see his challenge fulfilled. President Richard Nixon called the astronauts.
For every American, this has to be the proudest day of our lives.
It's an honor for us, Mr. President.
Armstrong and Aldrin picked up 47 pounds of moon rock and soil to bring back to earth. They set up experiments and took pictures.
I feel like a tourist, taking so many pictures.
Well, I suppose we are tourists.

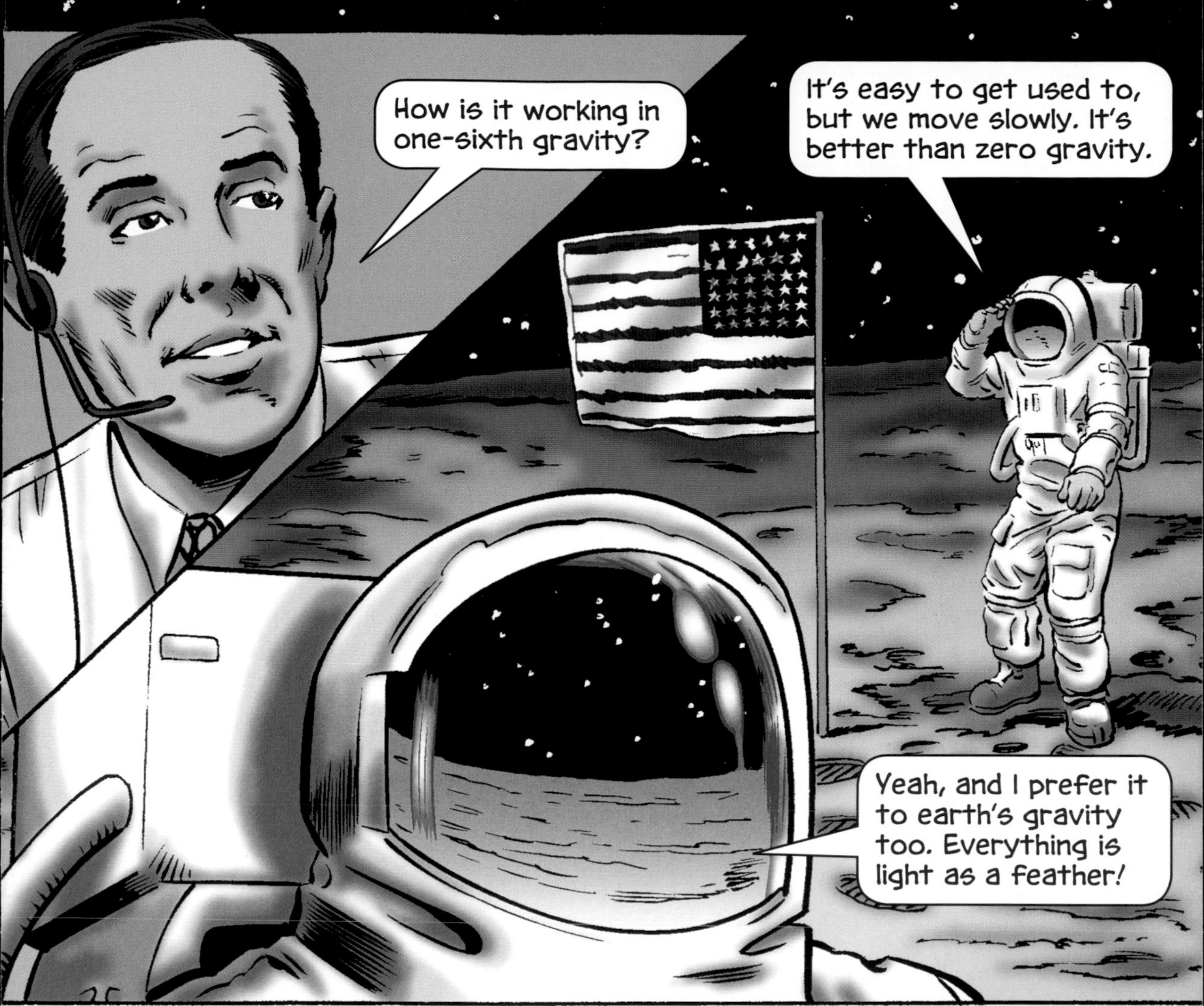
How is it working in one-sixth gravity?
It's easy to get used to, but we move slowly. It's better than zero gravity.
Yeah, and I prefer it to earth's gravity too. Everything is light as a feather!

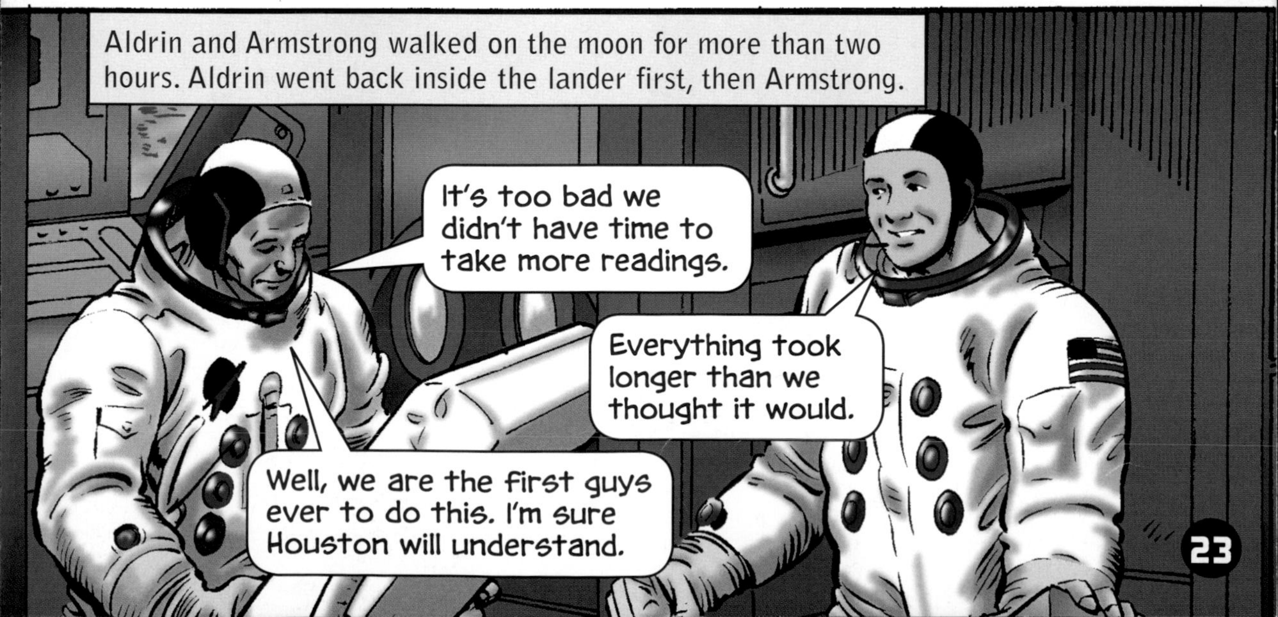
Aldrin and Armstrong walked on the moon for more than two hours. Aldrin went back inside the lander first, then Armstrong.
It's too bad we didn't have time to take more readings.
Everything took longer than we thought it would.
Well, we are the first guys ever to do this. I'm sure Houston will understand.

CHAPTER 4

NATIONAL HEROES

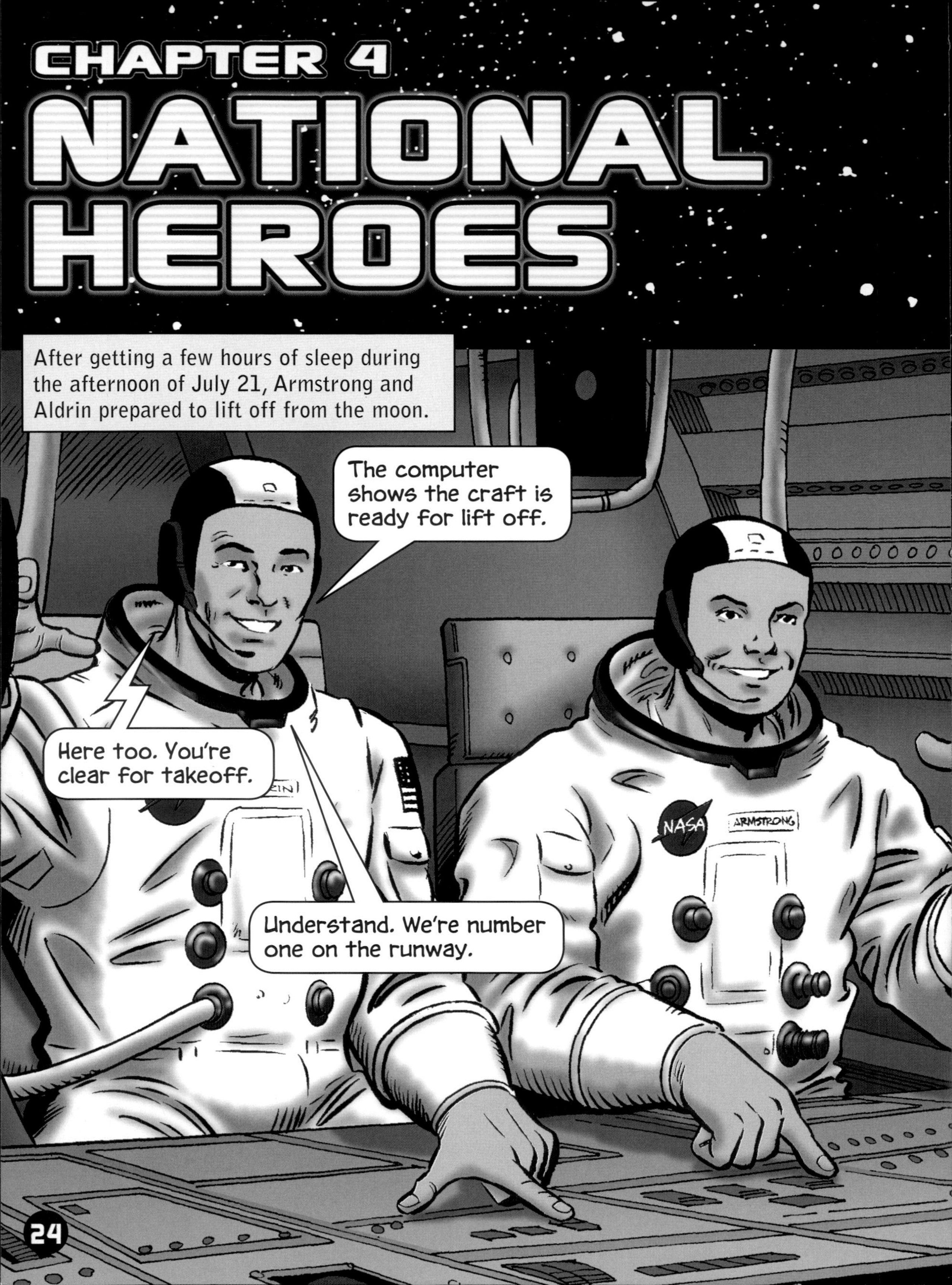
After getting a few hours of sleep during the afternoon of July 21, Armstrong and Aldrin prepared to lift off from the moon.
The computer shows the craft is ready for lift off.
Here too. You're clear for takeoff.
Understand. We're number one on the runway.
NASA
ARMSTRONG

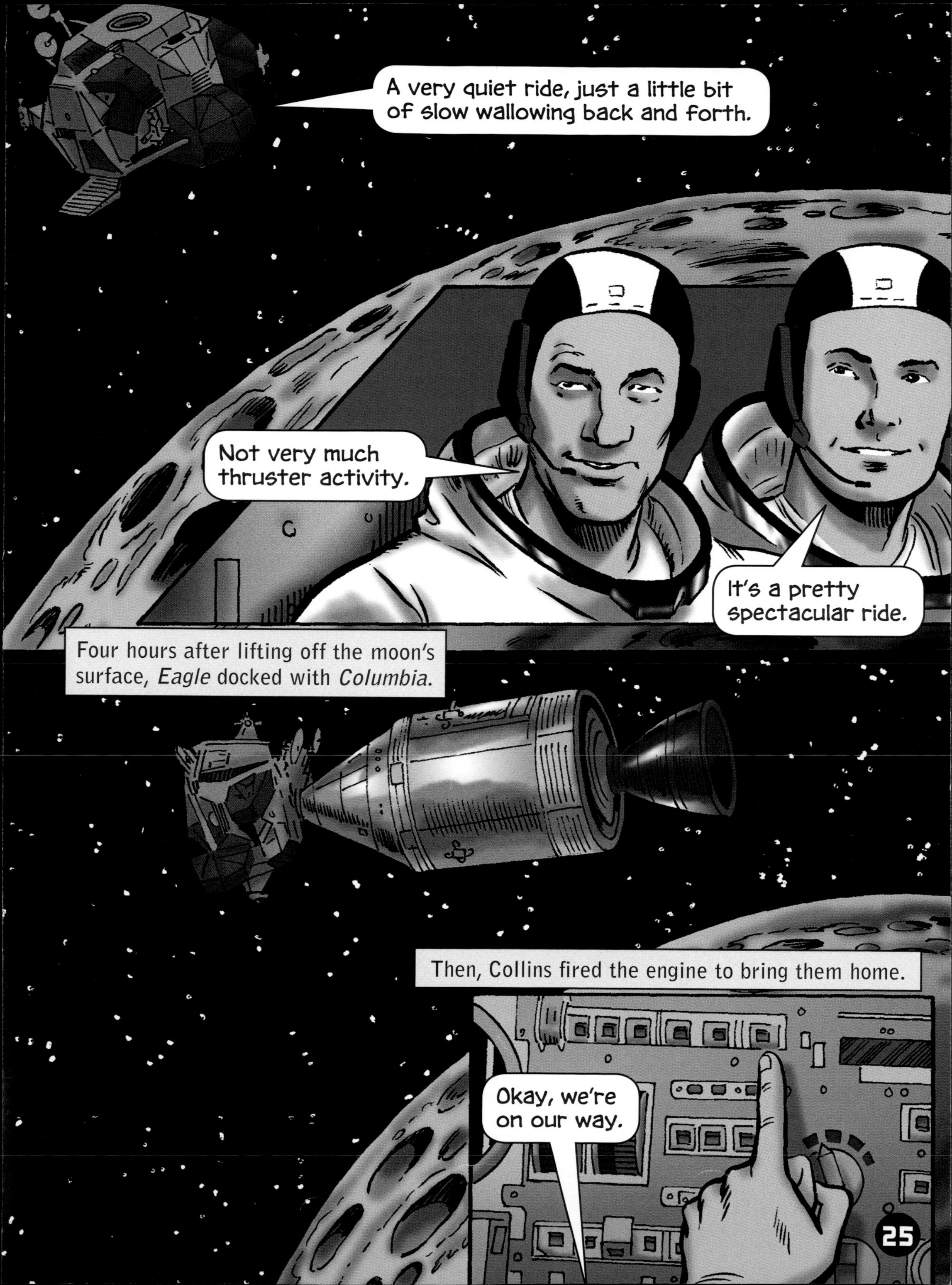
A very quiet ride, just a little bit of slow wallowing back and forth.
Not very much thruster activity.
It's a pretty spectacular ride.
Four hours after lifting off the moon's surface, *Eagle* docked with *Columbia*.
Then, Collins fired the engine to bring them home.
Okay, we're on our way.

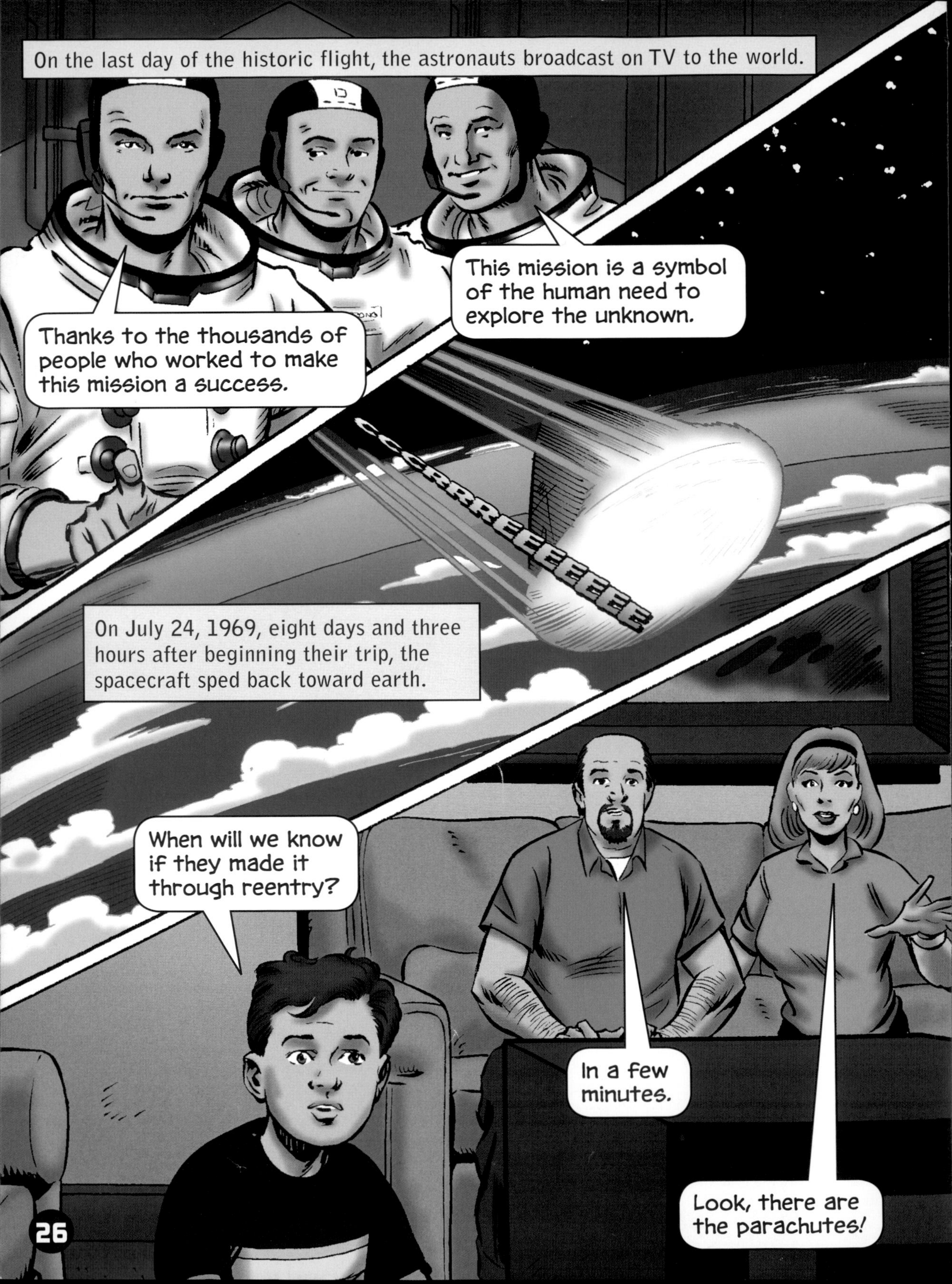

On the last day of the historic flight, the astronauts broadcast on TV to the world.
Thanks to the thousands of people who worked to make this mission a success.
This mission is a symbol of the human need to explore the unknown.
CCCRRREEEEEEEEE
On July 24, 1969, eight days and three hours after beginning their trip, the spacecraft sped back toward earth.
When will we know if they made it through reentry?
In a few minutes.
Look, there are the parachutes!

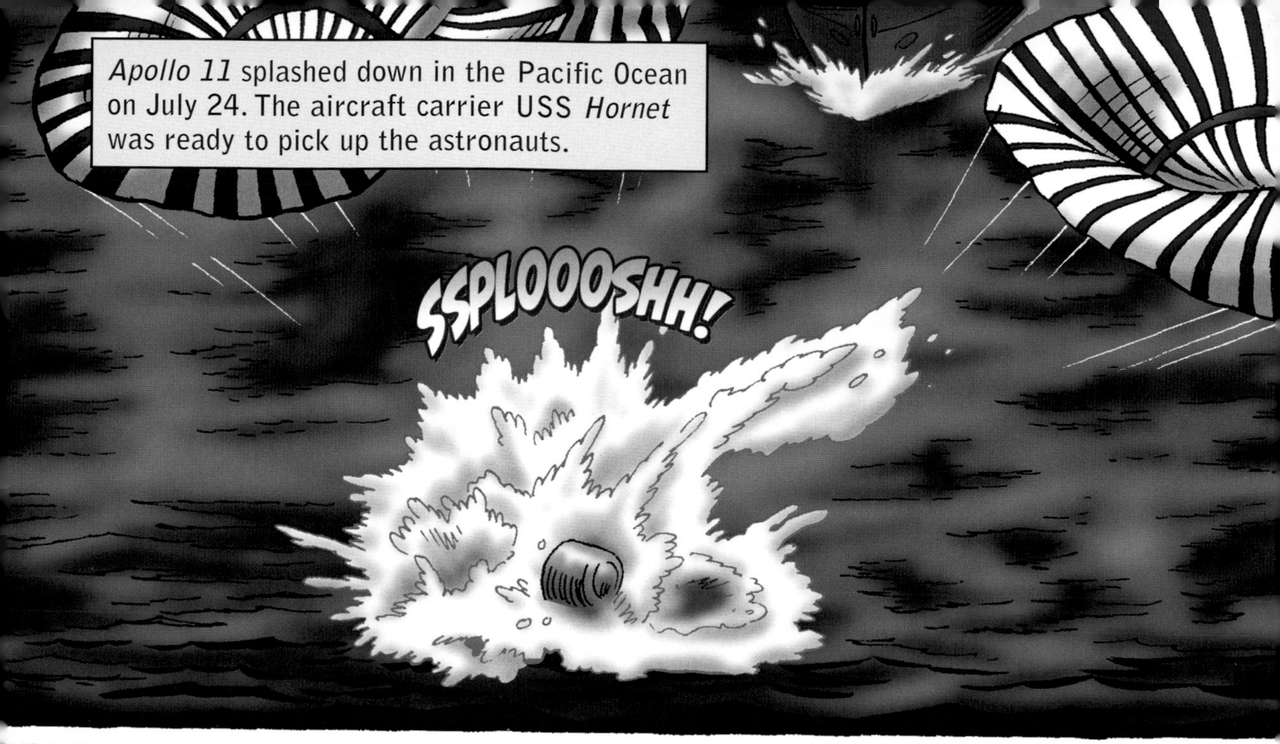
Apollo 11 splashed down in the Pacific Ocean on July 24. The aircraft carrier USS *Hornet* was ready to pick up the astronauts.
SSPLOOOSHH!

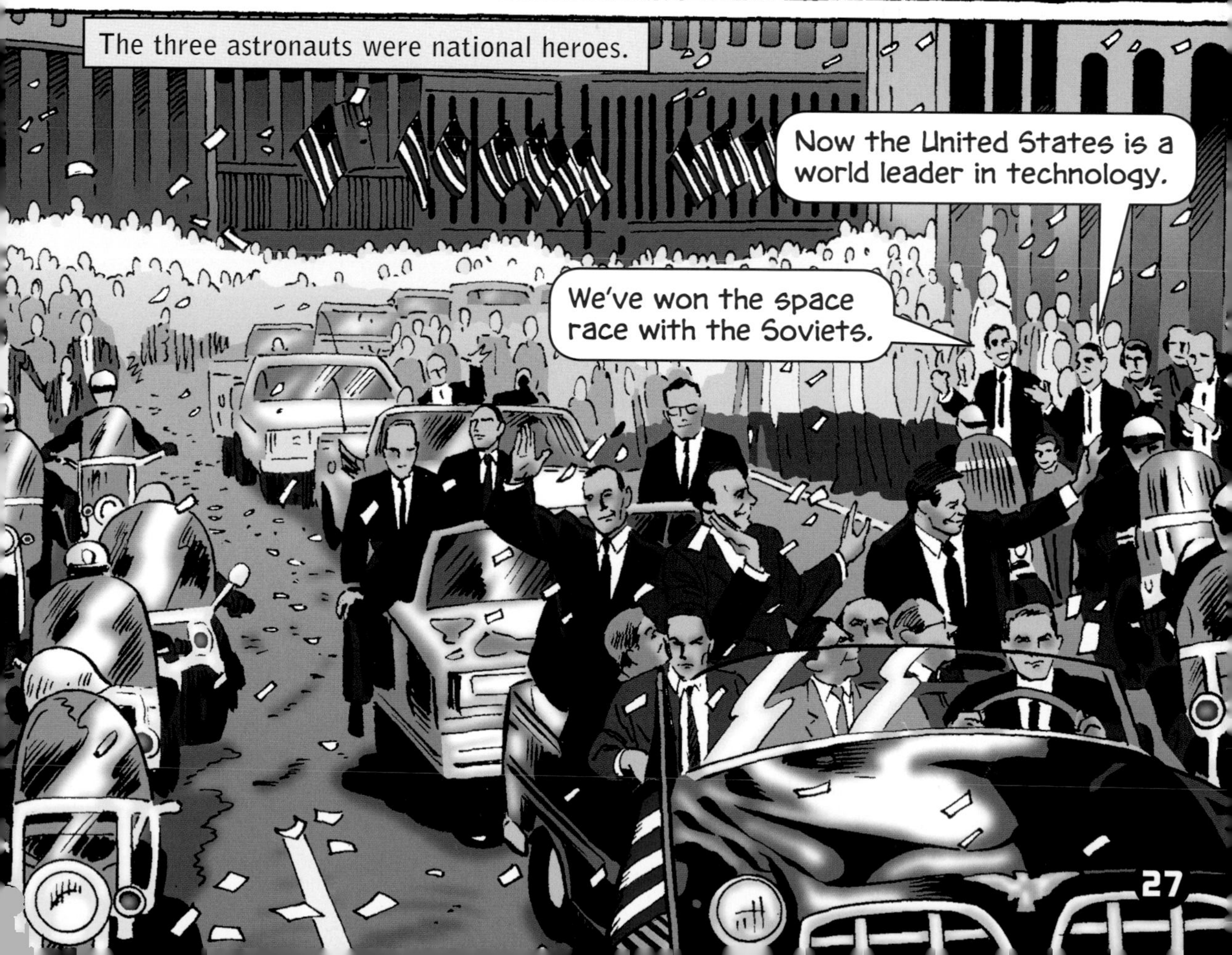
The three astronauts were national heroes.
Now the United States is a world leader in technology.
We've won the space race with the Soviets.

MORE ABOUT APOLLO 11

Near the end of their flight, the *Apollo 11* astronauts thanked the thousands of people involved in their historic mission. In fact, about 400,000 people working on the U.S. space program helped make the mission a success.

After stepping on the moon's surface, one of Armstrong's first tasks was to collect a "contingency sample." In case the astronauts had to lift off right away, he was to collect a small sample of moon soil. Armstrong quickly gathered three samples and placed them in his pocket.

One month before the *Apollo 11* liftoff, the Soviet Union attempted to launch an unmanned mission to the moon. The launch failed. Another attempt to collect samples was made on July 20, 1969. This unmanned spacecraft crashed on the moon.

Commander Neil Armstrong walked on the moon for 2 hours, 31 minutes, and 40 seconds.

From liftoff to splashdown, the *Apollo 11* mission lasted 8 days, 3 hours, 18 minutes, and 18 seconds. During that time, the astronauts traveled nearly 1 million miles.

- The astronauts left a plaque on the moon that reads:

HERE MEN FROM THE PLANET EARTH
FIRST SET FOOT UPON THE MOON
JULY 1969, A.D.
WE CAME IN PEACE FOR ALL MANKIND

- After the *Apollo 11* mission, NASA completed five more missions to different parts of the moon's surface. The last astronauts to walk on the moon were Eugene Cernan and Harrison Schmitt on December 14, 1972, during the *Apollo 17* mission.

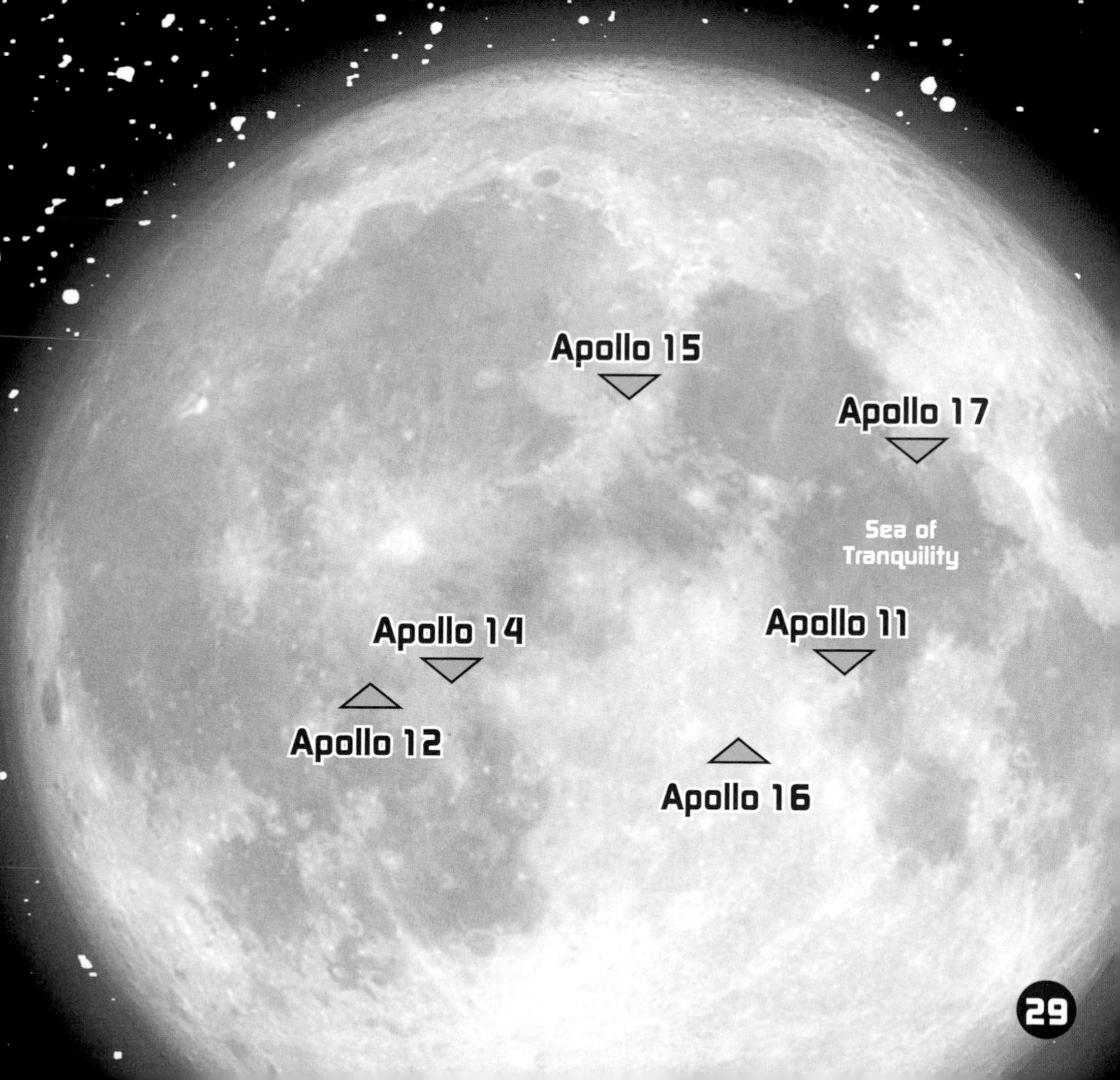

Apollo 15
Apollo 17
Sea of Tranquility
Apollo 14
Apollo 11
Apollo 12
Apollo 16

GLOSSARY

abort (uh-BORT)—to stop something from happening in the early stages

Congress (KONG-griss)—the government body of the United States that makes laws, made up of the Senate and the House of Representatives

maria (MAHR-ee-uh)—large dark areas on the moon caused by lava that flooded craters

module (MOJ-ool)—a separate section that can be linked to other parts

satellite (SAT-uh-lite)—a spacecraft that circles the earth; satellites gather and send information.

Soviet Union (SOH-vee-et YOON-yuhn)—a former federation of 15 republics that included Russia, Ukraine, and other nations in eastern Europe and northern Asia

INTERNET SITES

FactHound offers a safe, fun way to find Internet sites related to this book. All of the sites on FactHound have been researched by our staff.

Here's how:
1. Visit *www.facthound.com*
2. Choose your grade level.
3. Type in this book ID **073686492X** for age-appropriate sites. You may also browse subjects by clicking on letters, or by clicking on pictures and words.
4. Click on the **Fetch It** button.

FactHound will fetch the best sites for you!

READ MORE

Aldrin, Buzz. *Reaching for the Moon.* New York: HarperCollins, 2005.

Hudson-Goff, Elizabeth, and Dale Anderson. *The First Moon Landing.* Graphic Histories. Milwaukee: World Almanac Library, 2006.

Koestler-Grack, Rachel A. *Moon Landing.* American Moments. Edina, Minn.: Abdo, 2005.

Raum, Elizabeth. *Neil Armstrong.* American Lives. Chicago: Heinemann Library, 2006.

BIBLIOGRAPHY

Aldrin, Buzz, and Malcolm McConnell. *Men from Earth.* New York: Bantam, 1989.

Apollo 11 PAO Mission Commentary Transcript. http://www.jsc.nasa.gov/history/mission_trans/apollo11.htm

Armstrong, Neil. *First on the Moon: A Voyage with Neil Armstrong, Michael Collins, and Edwin E.* Aldrin, Jr. Boston: Little, Brown, 1970.

John F. Kennedy Library and Museum Online. http://www.jfklibrary.org/

Shepard, Alan, and Deke Slayton. *Moon Shot: The Inside Story of America's Race to the Moon.* Atlanta: Turner Publishing, 1994.

INDEX